Pride & Ownership

To Dad,

Be Safe Brother!

"Lutt"

1343

Pride & Ownership

A FIREFIGHTER'S LOVE OF THE JOB

Rick Lasky

> **Disclaimer**
>
> The recommendations, advice, descriptions, and the methods in this book are presented solely for educational purposes. The authors and publisher assume no liability whatsoever for any loss or damage that results from the use of any of the material in this book. Use of the material in this book is solely at the risk of the user.

Copyright © 2006 by
PennWell Corporation
1421 South Sheridan Road
Tulsa, Oklahoma 74112-6600 USA
800.752.9764
+1.918.831.9421
sales@pennwell.com
www.pennwellbooks.com
www.pennwell.com
www.fireengineeringbooks.com

Director: Mary McGee
Managing Editor: Jerry Naylis
Production/Operations Manager: Traci Huntsman
Production Editor: Amethyst Hensley
Senior Design Editor: Robin Remaley
Cover Designer: Clark Bell
Book Designer: Robin Remaley

Library of Congress Cataloging-in-Publication Data

Lasky, Rick.
   Pride & ownership : a firefighter's love of the job / by Rick Lasky.
     p. cm.
   Includes index.
   ISBN-13: 978-1-59370-078-2 (hardcover)
   ISBN-10: 1-59370-078-4 (hardcover)
   1. Fire fighters--Job satisfaction. 2. Fire fighters--Training of. 3. Fire departments--Personnel management. 4. Fire extinction--Vocational guidance. 5. Leadership. I. Title: Pride and ownership. II. Title.
TH9119.L365 2007
363.37023--dc22
                                                              2006008318

All rights reserved. No part of this book may be reproduced, stored in a retrieval system, or transcribed in any form or by any means, electronic or mechanical, including photocopying and recording, without the prior written permission of the publisher.

Printed in the United States of America

13 14 15 16 17 18 17 16 15 14

To my first family; Jami, Rick and Emily.
Thank you for always being there for me.
You are my life! To my mom, dad, and my
sister Doreene. Thanks for putting up with me
and for not kicking me out of the house when
you probably should have! Dad, thanks for
introducing me to the fire service.
And to my mother and father-in-law, Tonya
and Ascher Carl. Thanks for Jami.
I can't imagine life without her.

To my second family, the fire service.
Thank you for allowing me the privilege
and honor to serve with you.
You serve to everyone as the definition
of selflessness.

To our brothers and sisters that have given
their lives to protect the lives and property of
those you did not know and never met.
I will never forget you.

Never forgetting means never forgetting!

And to my younger brother Darren.
I wish you were here. I'm doing good buddy.

I love you and miss you!

Contents

Preface xiii
Acknowledgments xv

1 Our Mission 1
   Fires 1
   Emergency Medical Services 3
   Hazardous Materials 4
   Rescue 4
      You describe it and we do it! 7
   Weapons of Mass Destruction (WMD) 8
   Rubbing Two Nickels Together to Make a Quarter 8
      How much more can we do? 9
   Jack-of-All-Trades 9
   Where Did They Go? 10
      Fight the right fight 11
      Get off the dime 11
   Family Values 12
   Core Values 13

2 The Firefighter 15
   A Family of Very Special People 17
      The firefighter 18
      Integrity 20
   A Picture Says a Thousand Words 22
   The Denver Rescue 23
      Honor 24
      Pride 24
   My Saw 25
   Take Pictures 27
   Put the Firefighter Back in the Firefighter
   and the Firehouse Back in the Firehouse 28
      What kind of firefighter do you want to be? 31
   Do the Right Thing 31
      The rookie 33
   Go-to Guys 34
      Your own size-up 35

3 The Company Officer 37
A Very Proud Profession 37
Before They Walk Out the Door 38
We Used to Give Orders 38
Respect the Job and Each Other 39
It's a Privilege 42
    Share the history 42
Time for a History Lesson 42
    Bring back the mentors and train 43
Is it Worth the Work? 44
It's Worth the Work 44
Do We Really Get Anything Out of It? 45
When Will We See Results? 45
Where Do We Begin? 46
The Lewisville Program: Tapping into Years of Experience 46
The Five Questions 47
The Mentor Book 52
It Has to be Fair to All and Objective 52
From Firefighter to Driver Engineer 53
From Driver Engineer to Captain 54
From Captain to Battalion Chief 54
Division and Assistant Chiefs 54
Define Expectations 55
Firefighter Survival Training and Rapid Intervention Teams 55
    Fight for new equipment and be honest with what's killing us 59
Fireground—Battleground 59
    Our leadership 60
    Your circle of influence 61
Insist That They Appreciate This Job 63
    Learn to market your fire department 63
Get the Guys Out of the Firehouse 64
    Big Hat, No Cattle 65
The Neighborhood Wants a "Firehouse" Down the Block 65
    The best job in the world 65

4 The Chief 67
The Big Chair 68
    My favorite saying: egos eat brains 69
    Attitude 69
Lacking Foundation 70
Accountability 71
    Remember where you came from, Chief! 72
Good Communication = Good Labor Relations 72
    Learn about people 73
People Size-up 73
    Read smoke—read people 73
    Trust your people 74
The Little "Gold" Book 74

Put Some Good Stuff in their Files 75
    Let go of the past—focus on the future 75
    Build tomorrow's leaders and successors 76
    The chief's aide: A nearly extinct species making a comeback.................. 77
When Did We Lose Them? 78
Bringing a Species Back... 80
What's in a Name? 81
Some More Benefits 81
Since Implementation 83
The Information and Format is Out There... 84
They Get Promoted Fast 85
    This one's easy! 85
    Be careful what you wish for 86

5 Our Two Families 91
Defending Our Family 92
Don't Allow Anyone to Tarnish Our Image.. 92
Hiring the Right People 93
The I.A.F.F. and the F.O.O.L.S... 94
    Our *first* family 94
Setting Your Priorities 95
Consider Ride-outs... 95
Managing the Consequences of Other People's Actions............................... 98

6 Sweating the Small Stuff 99
The Line-of-Duty-Death Book Report Drill 100
The Little Things... 101
The Three "F's" 104
The First "F"—Our Firefighters 104
RIT Doesn't Suck! 106
Your People are a Reflection of Your Own Self Image 107
We're Losing on the Streets, Big Time 107
    Seat belts save lives 108
People Staging.. 109
The Second "F"—Our Fire Apparatus 110
The Third "F"—Our Firehouses 111
    Rule number 1 112
    Rule number 2 112
    Rule number 3 112
    Rule number 4 113

7 Changing Shirts: The Promotion 115
Don't Become Stagnant 116
The Decision to Promote 117
    First, why do I want to promote? 117
    Can I do the job? 117
    Can I make the life style change required? 118
    Do I have the courage to lead? 118

Study Habits 119
Provide a Good Process 121
Going to Days 121

8 What 9/11 Did To Us *And* For Us 125
The Public Outcry and Support 125
A Wake Up Call 126
Where Were the Experts Before 9/11? 126
    The funding is here! Well, some of it… 126
    Good politicians—bad politicians 128
    Just like the stop sign 129
We're Never Going to See a Disaster Like that Again 129
They Made a Difference, Again! 130

9 Ceremonies That Stoke the Flames of Tradition 131
There Is a Business Side to What We Do, But… 131
    The right tradition is not a bad thing 132
Ceremonies 132
    The new firefighter 133
    The promotion 137
    It's a big day for their family as well 138
    Retirements 138
    Graduations 142
    Awards ceremonies 142
    Wear them to honor those around us and our service 143
    New apparatus 145
    Our fireboat 146
    New firehouse 148
The 20-year Anniversary Firehouse Dinner 149
Be Careful, Because When They're Gone, They're Gone! 149
The Time to Honor Someone is When They are Still with Us 150

10 Marketing Your Fire Department 153
We Need to Market What We Do, All of It! 154
Marketing Our Mission 154
It's Not Really New 155
Let Us be the Shining Star for a While 155
    This stuff really works 156
Some Successful Programs 156
    "Vested for Life" program 156
    "Blazing a Trail for Literacy" program 158
    "After the Fire" program 160
    Our customer support unit 162
    Providing the foundation for stability 164
    "Santa Claus" program 164
    "Opening Day at School" program 165
    They just like helping people 173
    A "sod" story 174
    Helping someone get back on the right track 174
    If you would do it for your family, then do it! 175

Building Relationships 176
    That whole PD vs. FD thing again 176
    Working with city hall 177
    Work with who has the "pull" 177
    Seize the moment 178

11 Making It All Happen and Taking Care of Number One **179**
    Are You Being Honest with Yourself? 181
    Embracing Success 181
    Stop and Smell the Roses 182
    Learn from Your Successes the Same as You Do Your Failures 183

12 Have You Forgotten **185**
    Using 9/11 as a Crutch 185
    Never Forgetting Means Never Forgetting 187
    Forgetting Just One Is One Too Many 187
    Back to that Leadership Thing Again 188
    Don't Confuse a Tribute with Honoring Someone 192
    The Fire Service Is the Greatest Profession in the World 192

Appendix A: Mentoring **195**

Appendix B: Departmental Communications **243**

Appendix C: Ceremonies **251**

Appendix D: After the Fire **275**

Appendix E: Miscellaneous Forms **297**

Preface

The fire service is a very special organization and one that is second to none. As a matter of fact, there are many in the private sector, many corporations, who wish they could model themselves after it. It has been said for years that there are Fortune 500 companies that would kill for the marketing advantage the American fire service has, and that's mainly due to that fact that the public trusts us. They trust us with everything! At a time when the American family is struggling with divorce, abuse, and a lot of other problems, the fire service continues to serve as a role model for honesty. So many of our firefighters and officers have worked hard to get the fire service to where it is at today; for that very reason, we need to protect it. We're all about family. We're all about taking care of people. We're all about supporting and promoting family values. We need to protect what we're all about.

But for us to continue this journey well into the future and to protect it for future firefighters, we need to ask ourselves some questions. We need to be honest with ourselves.

- Who owns your fire department?
- It's been said for years that being a firefighter is the best job in the world. Why is it the best?
- What kind of a leader does it take to provide that feeling?
- Just as important: What kind of firefighter does it take?
- Do you own your fire department?
- Do you know where it all started and why?
- Do you have that pride, that love for the job, *or* do you just show up?

The reality is that some people are not cut out for the fire service. They just can't seem to make the commitment that is needed. But that's okay, because it's not for just anybody. It's for those who can commit to core values such as Pride, Honor, and Integrity. It's for those who can commit to a life of selflessness, to their brothers and sisters, and to the public. If this is you, welcome. I promise you it will be the best career choice you could have made. If you're already one of us and just need a shot in the arm to reenergize you or a "systems check" when it comes to you and the fire service, this book will move you in the right direction.

There's nothing else in the world like being a firefighter.

It is the best job in the world!

Acknowledgments

To my friends and mentors: There is no way to truly express my gratitude and thanks right here, right now, for everything you have all done for me. Each and every one of you has touched my life in some way. I feel blessed to have known each one of you and without you *Pride & Ownership* would not exist.

The PennWell and Fire Engineering staff, both past and present: Bob Halton, Bill Manning, Diane Feldman, Mary Jane Dittmar, Rob Maloney, Peter Hodge, Christine King, Tom Brennan, Glenn Corbett, Skip Coleman, Eric Schlett, Mary McGee, Julie Simmons, Francie Halcomb, Barbara McGee, Jerry Naylis, Jack Murphy and John Lewis.

The FDIC Firefighter Safety and Survival Team: The fire service is a safer place because of you!

Battalion Chief John Salka, FDNY: Thanks for your friendship, John!

Chief Steve Bass, Grapevine (TX) Fire Department
Deputy Chief Curtis Birt, Lake Cities (TX) Fire Department
Chief Alan Brunacini, Phoenix (AZ) Fire Department
Division Chief Eddie Buchanan, Hanover (VA)
    County Fire Department
Firefighter Seth Dale, Darien-Woodridge (IL) Fire District
Deputy District Chief Eddy Enright, Chicago (IL)
    Fire Department ret.
Chief Tom Freeman, Lisle-Woodridge (IL) Fire District.
    Love ya buddy!
Chief Kenny Gabriel, Coeur d'Alene (ID) Fire Department
Battalion Chief Billy Goldfeder, Loveland-Symmes (OH)
    Fire Department
Battalion Chief Don Hayde, FDNY
Battalion Chief John Hojek, Oak Lawn (IL) Fire Department
Chief Ron Kanterman, Merck (NJ) Fire Department
Chief Jack MacCastland, Illinois Fire Service Institute
Lieutenant Sal Marchese, FDNY ret.
Deputy Assistant Chief John Norman, FDNY

Captain Homer Robertson, Fort Worth (TX) Fire Department
Chief Bob Rubel, Bedford Park (IL) Fire Department ret.
Division Chief Tom Shervino, Oak Lawn (IL) Fire Department ret.
Chief Ron Szarzynski, Justice (IL) Fire Department ret.
Chief Dick Vachata, Pleasantview (IL) Fire District ret.

The Coeur d'Alene (ID) Fire Department
The Darien-Woodridge (IL) Fire District
The Illinois Fire Service Institute

Mike Gilbert and the F.O.O.L.S. To the brothers!

Some of the best photographers: Tony Greco, FDIC Photographer; Sidney Eads, Battalion Chief ret., Farmers Branch (TX) Fire Department and Lewisville FD Photographer; Tim Phillips, City of Lewisville Media Specialist

Head Athletic Trainer Dave Surprenant; Goalie Marty Turco; and the entire Dallas Stars Hockey Organization

John Travolta and the family at JTP Films

The City of Lewisville: Mayor and City Council; Human Resources Director Melinda Galler; and my bosses (the best I've ever worked for!), City Manager Claude King, Assistant City Manager Donna Barron, and Assistant City Manager Steve Bacchus

Mayor Steve Judy, City of Coeur D'Alene (ID) ret.

One of the best fire departments in the country, the men and women of the Lewisville (TX) Fire Department

Chief Ray Downey, Lieutenant Andy Fredericks, Lieutenant Billy McGinn and Firefighter Dana Hannon. I won't forget you!

Lieutenant Joe Samec. The reason there is "Saving Our Own."

1

OUR MISSION

For more than 200 years, our fire service has responded to just about every type of call for help and emergency imaginable without hesitation, attitude, or complaint. And over the years those calls for help have grown in number and complexity. The needs of our communities have pushed and prodded us into performing tasks and handling situations that no one ever imagined the fire service handling (fig. 1–1).

Fig. 1–1. The bell on this fire engine will long outlive the apparatus itself and will pass from generation to generation and from old engine to new.

Fires

Our mission as a fire service years ago was a simple one: to put out fires. Back then, the vision the public had of a firefighter was one that included everything from leather buckets, horses, bells, hoses, and fire axes, to the guys hanging around at the firehouse. But through the years the public has grown to understand that a firefighter will and can do just about anything, especially when it comes to helping you, your family, or your business (fig. 1–2).

We started out fighting fires, and in the years that followed, we took on one new area of responsibility after another and went from just horses and bells to a lot more (fig. 1–3, 1–4).

Fig. 1–2. The second motorized fire apparatus in Coeur d'Alene Idaho, this 1924 American LA France engine still proudly serves in parades and other department functions.

Fig. 1–3. There was a time "back in the olden days" when all we did was fight fires. Today most fire departments are "full service" organizations doing more than was ever imagined.

Fig. 1–4. Most fire service traditions are still alive and doing well and one that is coming back strong is the use of the 2½ inch attack line for fires.

Emergency Medical Services

It wasn't long before it was realized that we could provide first aid. That when you were hurt or hurting, to call us. We started with basic first aid, Red Cross first responders, and the like, and then moved into the emergency medical technician field—you know the guys with the big patches—which elevated the Emergency Medical Services (EMS) platform just a little bit higher. And then came paramedicine, thanks in part to the television show "Emergency." You know, the show with Squad 51 and L.A. County Firefighter Paramedics Johnny Gage and Roy Desoto. That show did more for the fire service and for the field of EMS than anyone could have imagined. "Emergency" won support from the public and brought a little more attention to what we do and to the fact that we don't just fight fires. It was also probably one of the best recruiting tools that accidentally fell into our laps. I can't tell you how many firefighters I know who say that show is the reason that they are in the fire service and is what made them want to be firefighters. And, to top it all off, Randy Mantooth, who played firefighter paramedic Johnny Gage, has over the years become a big fan of the fire service. His support for us has been nothing short of phenomenal. He has delivered keynote speeches, made guest appearances, and helped fire departments and fire associations with fundraising events across the United States of America. He's an awesome guy and the fire service is fortunate to have a fan in Hollywood like Randy Mantooth.

Hazardous Materials

Shortly after our introduction to paramedicine, we began getting called to chemical spills and releases, because, when it came down to it, no one else would respond or take care of them. Due to the serious nature of these incidents, the public called upon the fire department to handle them. So, we developed hazardous materials response procedures which have come a long way from washing it down the sewer and walking through it, to a much more sophisticated and proactive approach. We kiddingly refer to our hazardous-materials technicians as our "mop 'n' glow guys." But anyone who has been around an incident involving hazardous materials, especially a bad one, knows that our haz-mat technicians are definitely the people you want to rely on when the ethyl-methyl-bad-stuff ends up on the ground or in the air.

Rescue

Soon after that came dive and swift-water rescue. The Lewisville Fire Department, located in the Dallas-Fort Worth Metroplex in Texas, includes a great example of a fire department dive-rescue team. The team includes fifty-six divers, two dive-rescue units, two fire boats, and other various water craft (fig. 1–5, 1–6, 1–7, 1–8). Next, we entered the specialized rescue fields such as high-and low-angle rescue, confined-space rescue (we were already one up on this one because of our background in hazardous materials) and trench rescue (fig. 1–9). There was a time when dirt

Fig. 1–5. The Lewisville (TX) Fire Department Dive-Rescue Training Tank logo and mascot

was just dirt but now there are three different classes of soil, Class A, B, and C. We treat all of them like Class C because of the engulfing hazard and everything else that is involved when you're dealing with unstable ground. We moved into collapse rescue and continued to hone our skills with auto extrication. The biggest concern with auto extrications used to be the gas tank or battery cables, but now we have to worry about energy-absorbing bumpers blowing and hybrid cars and their associated dangers (fig. 1–10). We've found ourselves handling all types of rescues. And the advances that have been made in the area of technical rescue are absolutely amazing.

Fig. 1–6. The training tank dubbed the "Shark Tank" is located directly behind Firehouse #2, which serves as the main dive-rescue firehouse and could be the first of its kind in the country.

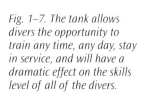

Fig. 1–7. The tank allows divers the opportunity to train any time, any day, stay in service, and will have a dramatic effect on the skills level of all of the divers.

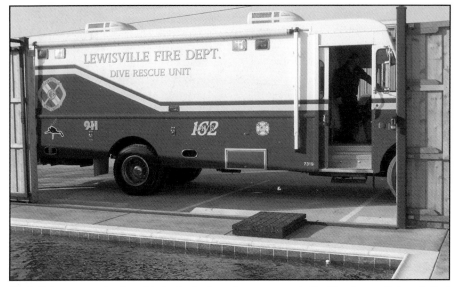

Fig. 1–8. An access gate alongside the tank allows for easy access to the Dive-Rescue Unit and all of its equipment.

Fig. 1–9. There used to be a time when dirt was just dirt, but the dangers associated with trench rescue emergencies require that the fire service provide quality training, equipment, and rescue teams.

Fig. 1–10. Today's car fires present a long list of dangers that were not seen twenty-five years ago.

Just look at the urban search and rescue (USAR) teams that Chief Ray Downey helped to create and how effective they are and what an important asset they have become (fig. 1–11). Take a look at any major disaster anywhere in the world and you'll see USAR teams from across the country, all made up of fire service people. This is mainly due to the fact that we have so many very talented people in the fire service and we know we can tap into their abilities, their talents, and their experiences. The end result when you do this is that everyone benefits, both the public *and* the fire service.

You describe it and we do it!

Over the years we became plumbers, electricians, roofers, and building construction and flood experts. A few years ago they turned us into hot water heater, stove, and furnace repairmen, this due in part to carbon monoxide detectors, commonly, but mistakenly, referred to by dispatch as "CO_2 alarms." Again, someone called us and the fire service responded, often doing so with little or no training and, at times, without the needed

Fig. 1–11 Chief Ray Downey

equipment, as was usually the case when the public threw us another new thing to handle. Specifically when we began responding to CO calls, we researched the problem, refined it, trained for it, wrote the standard operating procedures, and ended up with a standard that others have used as a model.

Weapons of Mass Destruction (WMD)

And just when you thought it couldn't get any worse, along came weapons of mass destruction (WMD). Before September 11, 2001, we talked about it and many fire departments trained for it. We discussed the idea of having to be trained for and responding to an incident involving WMD; many in the fire service thought it was nuts. They thought we were never going to have to deal with something like this because that type of thing only happens overseas or in some other country. On 9/11 we realized just how vulnerable we were to acts of terrorism and that our way of life had to change, both as civilians and as members of the fire service. I once heard a member of the military asked the question if he ever saw us getting back to normal, and his reply was, "We will get back to normal, but normal will have changed." Now, once again we have taken on another area of responsibility. We are at the forefront of Homeland Security. Most of our fire departments manage their local emergency management program because we're the ones who have planned for, prepared for, and responded to all major incidents and disasters. We've prepared our troops to march into the battle of homeland security issues and to handle everything from poison gases to major incidents of terrorism.

Rubbing Two Nickels Together to Make a Quarter

What makes it tougher is that we've had to do it with little or no funding. We've had to expand our programs and provide additional services without any additional money each time the fire service has been called upon to do something else beyond fighting fires, beyond what we always envisioned was the job of being a firefighter, beyond buckets, horses, bells, hoses, and axes.

How much more can we do?

Firefighters continually ask, "How much more stuff are they going to give us? They keep giving us more stuff to do! It seems that as soon as we learn a new area and skill, they throw something else at us." I've got news for you: it's going to keep happening as long as the fire service is as talented and as full of as many special people as it is. They are going to continue to call us every time there is a new problem or challenge. When you look at the history of the fire service as a whole, you'll see that, given any problem, we will always come up with a solution.

Jack-of-All-Trades

Firefighters truly are jacks-of-all-trades, but we are also masters of them all. Give us any challenge; we'll face it and overcome it. That is one of the most awesome things about the fire service. Call a firefighter, call the fire department, present them with a problem, and they will figure it out and work you through it. Remember Rubik's Cube? You give a citizen Rubik's Cube and they'll sit on the couch for weeks trying to get all of the colors matched up. Give it to a firefighter and what does he do? He peels off the stickers, puts them back in the right order, then throws it back to you and tells you to turn Jerry Springer back on! In fact, I've never seen people enjoy tackling challenges, solving problems, and helping people more than those within the fire service.

And it doesn't matter if a firefighter is paid or a volunteer; it's the same thing. We're there to help people. We're there to help families. Simply put, our best day is their worst day. When you're at your worst and life has you down, when you've got something horrible going on, whether it's a fire, medical problem, or some other type of disaster, we'll be there for you and help you, every time! We'll do everything we can to make things better again. That's what the fire service is all about. Firefighters are talented and are the cream of the crop in society. But, the same aggressiveness that makes us good at solving problems can get us into trouble, even to the point at which we lose a firefighter. We end up putting ourselves in a predicament that we shouldn't

"Our best day is their worst day."

be in. We go a step too far. We are so used to helping people, to stepping forward into harm's way to help our fellow man when others won't, that at times we take too many chances and we put our people in dangerous positions. Again, being aggressive in a lot of the things we do is one of our biggest attributes, but if taken too far it can get us into trouble.

Still, that's the nature of the beast in the fire service. It's not an excuse by any means, especially when someone gets hurt or killed, but it is something we need to be aware of if we are ever going to control it. What we as fire service leaders and future leaders must do is use every means possible, everything within our power, to make sure that we protect our personnel in every way imaginable. Whether it's protective clothing and assuring that they have the proper personal protective equipment (PPE), including good, reliable, and safe self-contained breathing apparatus (SCBA), assuring that they have portable radios and that they work, assuring that they have the training needed to do their job, the support of the fire department administration or the upper echelon, the proper apparatus, firehouses, tools and equipment, and, more than anything else, the proper amount of personnel. In short, make sure they have anything that will help them do their job better and stay safe.

Where did they go?

" ...back in the old days we did it with a lot fewer people."

It's kind of funny, before 9/11 there were many politicians, city managers and even fire chiefs beating up the fire service about staffing issues, about National Fire Protection Association (NFPA) 1710, saying "you don't need as many firefighters," "back in the old days we did it with a lot fewer people" and "it's just a push by the union for more people" and a lot of other nonsense. But after 9/11, you didn't hear much out of them. Well, let me back up just a little. You heard from them when they wanted to make an appearance with you or put their arm around you when the cameras were out or when they needed votes. They were the walls to climb and obstacles to overcome when it came to trying to increase staffing and obtain more funding. They didn't believe that we needed four people on a rig. No need, they said, and they told us to stop our whining. What amazes me is that these are the same people who won't go golfing without a foursome. They'll call everyone they know trying to get a fourth person to go golfing, but have no problem sending a two- or three-man company as a first-in company to a structure fire. And it's always staffing that they want to cut first. Our fight for staffing has to be strong and cannot end until we have the people we need and deserve.

Fight the right fight

We also need to fight continuously for better equipment, portable radios, training, facilities, and apparatus for our troops. There are a lot of fire departments out there who truly don't have the funding for more people and are fighting just to keep what they have and not lose anything. And they are pretty good departments. Now don't get me wrong; there are plenty of very good fire chiefs and some pretty good politicians. I'm just unloading on those who woke up one day and found out they were a chief or the mayor. I'm referring to those who have the money and don't want to spend it on the fire department.

Get off the dime!

The people who hold the purse strings need to realize that all we want to do is make it safer for our troops. They need to understand that all the unions really want is to assure that their members are safe and go home to their families. Pretty simple.

And where are those politicians now? It makes me sad and angry to read about another firehouse closing or the layoff of firefighters and to see that the politicians who don't understand the fire service can still find the money to pay themselves or take care of their pet projects. They also never seem to come up short when it comes to funding law enforcement. Yet they keep turning us down. Of course, that will change when we're

"A word of caution: be careful of who's behind you and what they're doing while they're there."

hit by the next "big one." Then they'll alternate between asking where we were and why weren't we prepared and having their pictures taken with us and patting us on the back. A word of caution: be careful of who's behind you and what they're doing while they're there. I try to keep politicians in front of me.

Our mission when you look at it is kind of split right down the middle. One is to provide the best protection possible for those we have sworn to serve and to provide a service to them that is second to none, and the second to promote family values. When the public needs us, no matter what the problem is, no matter what time of day it is, they can call on us. Even if it's out of our area of expertise, it doesn't matter, because we'll find someone who can help. If they didn't get their newspaper delivered; we'll give them the phone number that they need to call.

It has a lot to do with protecting our own and standing up for our profession and defending our heritage. We really don't need private firms or contractors out there doing our job for us. We don't need to privatize everything or outsource as much as we can. There are way too many people out there waiting and hiding and trying to take things away from us. We need to fight to hang on to what we already have and continue to fight for what we don't, but need.

Family Values

Our mission also has been to be there to promote family values. Whether you have a decent budget or not, one simple way to accomplish this mission is to promote family values. Treat people like family. Get your firefighters to treat the people they are responding to as if the people were their mom, dad, grandparent, or child (fig. 1–12).

Insist that they treat each structure as if it were their own home or business. If you can produce that kind of atmosphere, that kind of attitude, then all of the rest of it kind of falls into place. Your politicians will be happy. The boss will be happy. Your firefighters are going to enjoy the accolades for doing a good job and the thank you notes for going above and beyond the normal expectations of a firefighter. And most of the public, whom we have sworn to serve and protect, will be grateful for the services we provide. That's our mission. It's simple. Our mission is to take care of people.

Fig. 1–12. The fire service has and always will promote and defend strong family values. Assistant Chief Darrell Brown is pictured with three of his four daughters.

Core Values

We can't fulfill our mission and meet our goals if we don't live by core values. And I'm talking about realistic, attainable core values. We have to have something that the entire group believes in. We have to have something to stand by. We have to have something that people will see first hand that we stand for. My buddy John Salka and I have always wondered what they do to Marines in that short time of service that has them still wearing the jacket, shirt, or ball cap with the Marine Corps insignia on it even when they've been out of the Corps for years. The answer is an easy one: they instilled a set of core values that Marines live by, believe in, and support, not just for a while, but forever. There's a saying: "Once a Marine, always a Marine." It's very much the same for a firefighter. If you

> "Once a Marine, always a Marine."

have the love for the job, for your brothers and sisters, and for the honor of our profession, yes, you're at a point where you may feel that "once a firefighter, always a firefighter!" But without core values we have no vision, no guiding principles. Without a vision we have no way of fulfilling our mission. Without a mission we're like a bunch of ducks wandering in a thunderstorm, kind of hoping that what happens, happens for the right reasons.

We've learned from years and years of mistakes, failures, and successes that we have to plan our strategy. We have to have a plan of what we're going to do and have a plan to back that plan up. They dialed 911 for us. We can't dial 912. We're it! Bottom-line, our mission is to treat people like family and to take care of those around us. Our mission is very clear: to be there for people when they really need us.

> "Once a firefighter, always a firefighter."

2
THE FIREFIGHTER

What's amazing about the fire service is that when you look around you see the widest variety of people imaginable. Fire service members are young and old, experienced and green, and each has different interests (fig. 2–1). Some are interested in fighting fires, some want to be part of the emergency medical service, some are attracted to the specialty areas, and some are into all of it.

When *you* look at the fire service, what do *you* want to get out of it? Why did *you* get into it? Whether you're a paid firefighter, a volunteer, paid-on-call, or any combination, you obviously didn't get into it for the money. You're not going to get rich being a firefighter, which is the reason we all have side jobs. You joined the fire service to get something that money couldn't buy. You got into it to save lives and protect property, to help people, and to make a difference. You wanted to be part of a team that is second to none. Most of you can remember wanting to do this from the time you were little. I know I did, just as my son and daughter do now (figs. 2–2, 2–3, 2–4).

Fig. 2–1. As a firefighter you need to have a love for the job, the kind of love that never leaves you.

Fig. 2–2. My brother Darren (left) and I wanted to be firefighters from the day we were born.

I can remember hiding under my dad's turnout coat on the back seat of the car and scaring the you-know-what out of him when he got to the fire. He didn't know whether to leave me with the police or the driver of the pumper. He should have left me with the police; I probably would have stayed out of trouble that way.

Fig. 2–3. My son Rick has had the "itch" since he was very young.

A Family of Very Special People

The excitement of being a firefighter who responds to calls, fights fires, and helps people is unmatched by anything else. But once you got here, you realized there was more to it than just that. You found yourself in a whole new world. First of all, you realized that the fire service was made up of special people who value family. Little did you know that soon you'd belong to two families, one at home, and one down at the firehouse. You know the firehouse family I'm talking about. That's the one your spouse is jealous of. We've all heard, "He'd rather spend time at that firehouse than with me" or, as my wife would say, "In order for me to get his attention I need a siren on my head and red lights on my chest!" Well, come to think of it, that would grab one's attention. You know what I'm talking about; you're out there. My ex used to say "I can't get you off of the couch to go to my mothers, but that darn pager goes off and you're running down the driveway with one leg in your pants yelling, 'Yahoo!' What do I have to do to get you to do that when it's time to go to my mom's?" I said, "Simple. Start her house on fire."

Fig. 2–4. My daughter Emily is shooting for the top and already has a love for the fire service.

In all seriousness, it does take a special person to do this kind of work. It takes the kind of person who loves to help people and lives to be challenged. As I said earlier, give a firefighter any problem and he or she will solve it, give us any challenge, and we'll beat it. We'll do it and do it well, no matter what it is. Firefighters truly are jacks-of-all-trades, but we're also the masters of them all. We have to be, because that's what the public expects from us. We can figure out just about anything and that is a good thing, but we need to remember that at times it can also be a bad thing that can get us into trouble when we're a bit too aggressive. Even so, firefighters are talented and are the cream of the crop in society.

The firefighter

Let's talk about this person, the firefighter, for a minute. What qualities does it take to make a good firefighter? We already know that they need to care about people and love a challenge, but perhaps most of all they need to have a love for the job. They need the unshakable passion of being a firefighter, the one love that never leaves you. But how do they get it? I love this job more than life itself. My family comes first, but I love this job. That passion didn't just appear one day. It came from inside. It came from my father, who was a firefighter, and from being around great people like my mentors. All of them loved this job so much you couldn't help but admire them.

My love for the job came from people like Chief Jack MacCastland, who taught my Firefighter 1 class when I was 18 years old (fig. 2–5). Mac said then, "You want to be a good firefighter, you need to know building construction and fire behavior. You have to know how the building is going to react with the fire and how the fire is going to react with the building (fig. 2–6). Anyone can go out there and chop, but you still need to know building construction and fire behavior." How very true. The first time I saw Mac was at a fire. Here was this big guy, leather helmet all bent up in the front, pulling sheetrock with a pike pole in one hand and using his chin to open and close the nozzle in the other hand. I said, "Man, I want to be like this guy."

Fig. 2–5. Chief Jack MacCastland

Fig. 2–6. Mac always said, "You want to be a good firefighter, you have to know fire behavior and building construction. You have to know how the building's going to react to the fire and how the fire's going to react to the building."

Then there was Chief Eddy Enright, who taught me about caring for and respecting your guys (fig. 2–7). He'd always say, "Look for their positives, Rick. Anyone can pick out the negatives. You do that and the negatives tend to go away. Catch them doing something right. It's easy to catch them doing something wrong."

Fig. 2–7. Chief Eddy Enright

One of my best friends, Chief Tom Freeman, one of the smartest firefighters I've ever met, also shared his knowledge with me (fig. 2–8). He said things like "A good officer or incident commander is the one that can predict his next alarm. Any mope in a white helmet can stand outside and handle what he's got right now and burn it to the ground, but it's the guy that can predict his next alarm, knows when he needs more resources before he runs out, where the fire's going, and can think out of that box, that does well."

Perhaps the most important thing my mentors taught me was that to be a good firefighter you have to have *core values* such as *pride*, *honor*, and *integrity*. And every last bit of it starts with integrity. Unfortunately, there are a lot of people who profess to have integrity, but do not.

Fig. 2–8. Chief Tom Freeman

> "Pride, Honor, and Integrity."

Integrity

Your character is built on integrity. It's built on honesty with yourself and those around you. Having integrity takes having values. There's more to it than just saying that you have integrity. If I sleep in the garage one night, that doesn't make me a car. To truly have integrity you have to live your life with integrity. You have to say what you mean and mean what you say. If you lie, steal, or cheat, you have no integrity. Having integrity means that you do the right thing. Integrity is the foundation for character and your character is defined by how you act and what you do when no one is looking.

The public's trust in the fire service is nothing short of amazing. But we've earned it. They let us into their homes, their businesses (we've got all of the keys), call on us to take care of their loved ones, and trust us with just about everything. A nice example of this occurred the first week that my family and I were in Coeur d'Alene, Idaho. I had taken the chief's job there and had just finished my first week. When I got home and walked in the door my wife said, "You're never going to believe this one." She told me that a little bit earlier there had been a knock at the door. When my wife answered she was greeted by a lady who asked her if she was the fire chief's wife. After my wife replied that she was, the lady handed over the keys to her house and said, "I'm Nancy from down the street and we're headed out of town for the weekend and we were wondering if you would watch our house for us?" We had never met her before and had only been in town for a couple of weeks, but she felt comfortable enough and trusted us enough to ask us to watch her home because I was a firefighter.

If you want to see an example of someone with integrity, look to Bill Manning, the former chief editor of *Fire Engineering Magazine* (fig. 2–9). Bill worked for the fire service every single day, bringing issues to light and fighting for us when someone had wronged us. His editorials were great. Yes, some of them ticked people off, but they were supposed to! His editorials were supposed to get people off of their "sector C" and into the forefront. Remember the "White" magazine cover on the August of '95 issue? The one that was supposed to depict the perfectly safe fireground? There was no picture (fig. 2–10). That was Bill's response to all of the complaints the magazine received when their cover pictures showed firefighters who weren't following all safety rules, such as keeping their hoods on or having their boots pulled up.

> "Just because I sleep in the garage one night, that doesn't make me a car."

Fig. 2–9. Chief Editor Bill Manning

Fig. 2–10. The "White Cover" was Bill Manning's attempt to depict the perfect fireground in response to negative comments about some of the pictures used on the covers of Fire Engineering *magazine*.

A Picture Says a Thousand Words

Bill by no means condoned unsafe acts. His point was that you should look at the covers to see what's going on in the country and, more importantly, to compare it to your own department. Ask yourself what's going on in the picture and how your department would have handled it. What would you do? He wanted us to learn something from each cover. I've sat down at the kitchen table at the firehouse with one of those covers and gotten an hour's worth of tactics and strategy out of it.

The Denver Rescue

Did you see the April issue back in '93 that had the photo of a Denver Colorado firefighter unconscious in a room (fig. 2–11)? That one generated a lot of talk and some people were upset by it. They said, "How could you put a picture of a brother on the cover like that?" What they didn't realize was that the cover photo was a reenactment of the fire in Denver that claimed the life of veteran Engineer Mark Langvardt. The point being made by then Captain Dave McGrail, now District Chief McGrail (fig. 2–12), was that as hard as they tried, and they tried to the point of exhaustion, they couldn't get Mark out the window of this 6-foot by 11-foot room with only 28 inches of working space, a window sill height of 42 inches, and a window opening of 20 inches. They tried just about everything. Those of you who have drilled and trained on the "Denver Rescue" know just how difficult that rescue can be.

Fig. 2–11. This April 1993 issue of Fire Engineering magazine got the fire service talking about firefighter survival training.

The editorials and pictures are supposed to stimulate interest. They're supposed to get you thinking. A picture and an article can make a big difference. By sharing the details regarding the tragic death of engineer Mark Langvardt, Dave McGrail has literally saved the lives of hundreds of firefighters. He and his guys didn't try to hide it. They gave us the opportunity to see how we would react to this situation and what it takes to get a brother or sister out of it. For this we are extremely grateful. Thanks, Dave!

Fig. 2–12. Chief Dave McGrail

Honor

Honor is built through respect and loyalty. It means showing esprit' de corps and caring enough about those around you that you would do anything for them, on duty and off. Honor is the brotherhood. To me the brotherhood means more than just a sticker on the windshield of your car. It means that when your kids are sick we help out. That when you're having a tough time with your bills, we chip in. That when you need to move into your new house, we move you, and when that new house needs a new roof, we tear off the old one and we re-roof it. It also means that I would lie next to you and burn the ears off of my head before I would ever leave you in a burning building.

Honor is also preventing anyone from giving your company or department a black eye or doing anything to hurt its reputation. Peer pressure can really help when it comes to making sure people do the right thing. To see an example of honor, take a look at the majority of the instructors teaching in the fire service today. Most are trying to share information to make the fire service better or safer and want to do nothing more than make a difference in the lives of the firefighters they teach. Each is on a mission to teach firefighters how to go home from fires. I guarantee that you'll see honor, and pride, too.

> "Esprit de corps: feeling of loyalty to a group."

Pride

But pride doesn't just happen. Pride takes work. It requires ownership. I received my first true lesson in pride and ownership about 20 years ago. Tommy usually gets embarrassed when I tell this story, but too bad, Tommy. It's a great story.

My Saw

We were working a fire in an old school building. We were on the second floor, chasing fire in the void spaces, cutting floor away, and opening up walls. I noticed a crew across the room trying to get their saw started. This went on for a while. A couple of them put their axes down to help start the saw, forgetting that their axes will always do the one thing their saw won't: start. While this was going on, another company officer, a Lieutenant named Tom Shervino (fig. 2–13), looked at his chief and said, "Let me go get my saw, chief." The chief said to wait a minute longer but Tommy persisted. Finally his chief gave in and said, "Go get your saw, Tom." So off Tom went, soon returning with the saw. One pull started it and he began cutting. A short time later he stopped, went into the hallway, refueled the saw, and then went back to cutting. He had known when the saw was running low on fuel. He knew how to start it. He knew everything about it. The other crew never got their saw started, by the way.

At first I thought, how arrogant: *my* saw! What's with this *my* saw thing? Later, when I was outside getting ready to pick up and return to quarters, I saw Tom and asked him what he meant when he said it was *his* saw. He looked at me a bit confused—actually looked at me as if I had left my brain at home—and said, "That's not my saw. That's Oak Lawn's saw. But that's my squad so it's my saw today. That's my company." The tools he used weren't his personal items but he felt that he owned them that day, on his shift. Then it hit me. This guy was proud of his department, proud of his company, and proud of his tools, and with his pride came *ownership*.

Fig. 2–13. Chief Tom Shervino

There have been people lately who have written that "pride" is a bad thing. But I'm not talking about the pride that is associated with arrogance and creates problems. I'm talking about that feeling you get at the end of a job well done and the way you feel when you talk about your department, our heritage, and our traditions. That kind of pride is a *good* thing.

One way to foster pride, honor, and integrity is to hold promotional ceremonies. Have firefighters raise their right hand and swear to protect and serve. Have the Mayor or a family member who is a retired or active firefighter pin their badge on them (fig. 2–14). Let the important people in their lives join them on this special day and let them pin their collar pins on. This could be a spouse, one of their parents, a child or an officer who helped or mentored them. Make it special. Too many firefighters have gotten their badge in their mailbox or heard "give me that one and I'll give you this one" when they were promoted (fig. 2–15). Take lots of pictures and make it special, because it is special.

Fig. 2–14. Having the Mayor or a family member who is an active or retired firefighter pin their badge on them adds to a very proud moment.

Fig. 2–15. Having them raise their hand and swear to serve with their family present makes an already important event just that much more special.

Take Pictures

By the way, what has happened to the pictures of the guys that used to hang in the firehouse? You know, the pictures taken after a call or the ones of the guys standing next to their rig. My friend Peter Hodge and I were talking over a cup of coffee one day and he said, "You know we used to have the guys in the pictures with their rigs (fig. 2–16). Now we just have pictures of the rigs."

Another way to make your firefighters proud is to give them an awards ceremony every year. Invite their families. Have the victims from some of the incidents your department responded to present the awards if you can. This can really work out great for you. During one of our recent awards ceremonies, we gave out the Award of Valor and several other awards. One award was for a CPR save. While we had the companies on stage we invited the man who had been rescued to join us and present the awards to the guys who saved his life. Until that point nobody knew who he was. Most thought he was a relative of one of our firefighters or maybe a past politician. No one realized that he was the guy who had gone down in full cardiac arrest after jogging, there to thank them for saving his life. When we were done there wasn't a dry eye in the house. Celebrate your department's accomplishments and brag about them.

Fig. 2–16. Taking a group picture with your company after a good "job" used to be commonplace, but over the years this tradition has slid out of sight.

Put the Firefighter Back in the Firefighter and the Firehouse Back in the Firehouse

When you place a new rig in service, do it right. Use the radio to announce the retirement of the old rig and welcome aboard the new one, then give the rig a bath and push it into the firehouse (with the help of the driver) like we did in the "olden" days (fig. 2–17, 2–18).

Let each company have its own company logo. A logo is something to be proud of. When I got to Lewisville there were only two companies with logos. I asked the troops where the rest of the company logos were and they said there hadn't been much support from the administration in the past. Companies had to pay for them on their own (fig. 2–19, 2–20, 2–21, 2–22, 2–23, 2–24).

Fig. 2–17. In this picture the firefighters and citizens from the neighborhood participate in "pushing" the old rig out.

Fig. 2–18. As exciting as it is to push the old rig out, there is even more excitement when it comes time to push the new rig in. Both the firefighters and the citizens feel like they have ownership by doing this.

Fig. 2–19. Lewisville Firehouse #1 logo, "The Big House," named for the quantity of apparatus and the number of chief officers housed there.

Fig. 2–20. Lewisville Firehouse #2 logo, "Old Town Fighting 2nd," touting the local high school mascot "Fighting Farmer John" as a firefighter.

Fig. 2–21. Lewisville Firehouse #3 logo, "Raging Red," shows the bull on the west side, a very busy house.

Fig. 2–22. Lewisville Firehouse #4 logo, "The Fighting Mudcats," represents divers and the crew that operates the fireboats on Lake Lewisville.

Fig. 2–23. Lewisville Firehouse #5 logo, "Southside Border Crew," named due to the amount of automatic aid assistance they provide.

Fig. 2–24. Lewisville Firehouse #6 logo, "The Dragon Slayers," reflects a large residential subdivision within their district called Castle Hills. Homes there are built to look like the castles in Europe.

How much do decals and flags cost? How much of a return are you going to get back in pride (fig. 2–25, 2–26)? Support logos, support pride, and maybe you won't have to talk to them about taking care of the rig so often. Allow them to love the job and to have the passion.

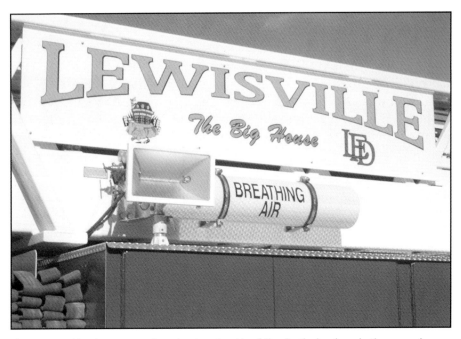

Fig. 2–25. Besides the company logo decal on the side of the rig, the logo's main theme can be placed on the main ladder of the truck company. Using the mascot or a specific symbol from the logo adds to company pride when the ladder goes up.

Fig. 2–26. The logo doesn't have to relate to a specific company but could represent a unit such as the Pipes and Drums.

What kind of a firefighter do you want to be?

With all of that in mind, you need to ask yourself this question: what kind of firefighter do you want to be? How do you want the people in your life and your fire service brothers and sisters to feel about you? Once again, what did you get into this profession for? What do you really want to get out of it? What you get out of the fire service starts on day one. Your first day!

For years, a lot of our folks have sat around and reminisced about how it used to be and told stories about that "star" rookie who walked into the firehouse, great attitude and all. They talk about the kind of new firefighter who, after you filled him or her in on how it all works and what is expected of a firefighter, had only one request: that when you needed something done, you think of them first. And many of these same people ask, "Where did they all go?" and complain that "You just don't see that kind of attitude anymore." The answer is a simple one. Those stars are right in front of us. They're the new guys walking in the door and most of the people we already have. It's just that we've forgotten to instill in them those core values we talked about.

If you want to be that star firefighter, do you have a passion for the job? If not yet, do you feel it growing inside of you? Are you willing to commit to our family and to our traditions? Understand that you really have to live by those core values. You can't be disrespectful of the fire service family. You can't steal from a brother or a sister or cut a brother's throat at the kitchen table with trash talk. You must be willing to do anything to make our profession better and to protect our family. You must be willing to do the right thing.

Do the Right Thing

Keep in mind that if what you are doing is to be the right thing, it has to be done for the right reasons. It cannot b the right thing if it's done in order to hurt someone. Do the right thing because it *is* the right thing to do! When you think about the things you did or said today, do you regret any of it? If you do, how can you fix it? Remember, what you say in a moment of passion can hang around for three years or longer. I remember being a fairly new firefighter and being assigned to A shift. After a while the

> "No matter what happened to you yesterday, your attitude is your choice today."

chief moved me from A shift to B shift. After all of the hard work we had put into our team and our routines, I was pretty upset. I thought B shift was where all of the misfits were and kept thinking that I was so much better than that, that I didn't really belong there, and that this was just not right. "Why me?" I thought. As you can imagine, my attitude was a little obvious. It didn't take long for me to alienate the entire firehouse and everyone on that shift, for that matter. It was at this point that my chief, a good man, called me into his office. He said "Don't sit down, you're not going to be in here that long," and proceeded to rip me a new you-know-what. He explained that he had moved me because he needed a little more strength on B shift. It was a way of spreading out the team players to make for a better overall operation and I had screwed up that part of it. He had thought I was smart enough to recognize what he was attempting to do, but I wasn't. He also explained that I'd had exactly the opposite effect I was supposed to and that it was going to take me a long time to reverse it all and gain back the confidence of my shift mates.

He was right. It took me quite a while. Realizing that I was wrong, big time, I left his office, walked right into the day room, and proceeded to apologize. I asked for another chance and said that I would prove to them that I was worthy of being on their team. I started doing extra work, helping my shift mates whenever I could, cooking the meals, and doing whatever I could to get back in their good graces. As the months went by, things got a lot better. About two years later, while I was working on a project with one of my shift mates, he explained how much nicer it was to work with me and said that I was okay after all. I was finally almost there, but look at all of the time I wasted having to fix something that I screwed up by thinking that I was better than my brothers and sisters and by being a little bit full of myself. The reality was that this shift was pretty good at what they did and that the shift I came from had actually needed a little help. I learned to keep an open mind, to look at the big picture, and to keep my mouth shut. I also learned not to fall into the trap of thinking that the shift before you are the slobs and the shift after you are the nitpickers. Come to find out, when you switch shifts, that your guys were the slobs or the nitpickers, and so on.

The rookie

As a rookie, remember you're starting your legacy when you walk through that door for the first time (fig. 2–27). Great firefighters have to come from somewhere. Why can't you be one? It really doesn't take much if you do your best to live by those core values that we discussed and remember that nobody's perfect. I've said for years that "perfect" people and know-it-alls in the fire service can get you hurt and killed. You don't have to be perfect to be a great firefighter. You just need good heart!

You need to do three things to be successful as a firefighter:

1) Do what is right

2) Do your best

3) Treat others as you would like to be treated . . . and then watch those relationships grow!

Fig. 2–27. As a rookie, you start your own legacy the moment you walk into the firehouse for the first time. What kind of a first impression do you want to make?

You can help yourself by getting involved with your department. Study it. Understand its heritage—where it all started and who got it there—and be concerned with where it is going. Soak up as much as you can. Get as much training as you can. Go to classes and attend seminars. Ask questions and read something about this job every day. Become as informed as you possibly can about our job because when you reach the point in your career when you think you know it all, you're a step away from disaster. Be the best you can be and stay positive. It's easy to sit around the kitchen table and find things wrong or complain. It's harder to think and talk about the good things.

Go-to Guys

Consider becoming one of the "go-to guys," someone who people can depend on to get the job done. Be the person the boss can go to for a job because they know you'll do it and do it well and usually without complaint. I don't mean that you should be a "sector C" kisser or apple polisher. Just be the reliable "go-to guy."

Fig. 2–28. Be proud of your uniform and of what it stands for.

Your own size-up

In closing here are some points to ponder:

- Show up on time for your shift or your weekly meeting or drill.
- Wear your uniform, take care of it, and be proud of it (fig. 2–28). As a friend of mine said, "If you worked for Orkin killing bugs, you'd have to wear a uniform." So why not for us?
- Take care of the firehouse. It is your home!
- Take care of your apparatus. It's not just an image thing. (Think about it!)
- What shape are your tools in? Are they rusted and stuck to the side of the rig?
- Are you still learning your trade? Remember to learn your trade before you learn the tricks.
- Train as if your life depends on it, because it does!
- Train on firefighter survival. Rapid intervention training (RIT) does work! Do you visit Chief Billy Goldfeder's web site, FirefighterCloseCalls.com, often or do you get his email "Secret List?" (Billy's a great fire service brother by the way!)
- Are you *controlling* or *contributing to* the gossip, rumor, and character assassination mill?
- Be willing to learn and never fall into the "I know everything" trap. Know-it-alls and "perfect" people will get you hurt or killed.
- Remember that you are creating your own legacy. It starts on your first day, the moment you walk into the firehouse.
- Take a good look at past firefighter fatalities and how they happened. What do we have to do to prevent them from happening again and what would be your role in the prevention?
- Find a mentor or several that will have a positive effect on you and on what you do.

- Defend our profession.
- Know our (the fire service's) history and your department's as well.
- Remember what you owe the public. Treat people like family and believe in and promote strong family values.
- Are you as proud of the job when off duty as you are on duty and vice versa? Promote a positive image and remember to keep the first job the first job and the second job the second job.
- Practice, preach, and promote safety.
- Remember what you say in three seconds of a heated moment can haunt you for three years. There are usually three sides to every story.
- Are you passionate about our profession and are you willing to leave it better for the next guy?
- Are we taking care of each other? For real?
- Everyone has a leadership role of some kind. Remember that it takes courage to lead. Make the right choices and decisions even though they may not be the most popular at the time. Go with what's right and honest. Integrity is where it all starts. Don't compromise your integrity for anyone.
- Every now and then, remind yourself why you got into this profession in the first place.
- Have fun! It's the best job in the world.
- Do you have a love for the job?
- And lastly, NEVER FORGETTING MEANS NEVER FORGETTING!

These are just a few suggestions and there are a lot more. We do have the ability to make a difference and start someone off on the right path and off to a great career in the fire service. You can make a difference if you want to. It's up to you!

3

THE COMPANY OFFICER

When you take an honest look at how the fire service works, the following things are obvious. First, our most important asset is our firefighters. Second, our battalion chiefs or shift commanders are the coaches. Third, it is the company officers, whether you call them lieutenant, sergeant, or captain, who gets things done and sets the tempo for the shift. As a chief officer you know that if you want to get things done you have to get to the company officer. This is both true and obvious when everything is going well, but it is also true when things are not going well. It all comes down to leadership and the company officer's ability to lead his or her troops in a good direction. But doing that also means taking on responsibility for yourself, for your actions, and for those you are going to lead.

A Very Proud Profession

The fire service has always been an extremely proud profession. But the past years have seen that pride begin to slip, and in some cases, to erode. Today we hear some of our officers and firefighters saying things like, "Why don't these guys care? They don't care about how the firehouse looks. They don't care about their uniforms. They don't take care of the rig. And they don't care about the job." And often they say, "There's nothing you can do about it. It's that generation X or Y. You know, the dot-com firefighters." You hear some officers and firefighters blame their parents. What they're saying is partially true, so isn't it time that we sit them down and tell them how it is in our world?

> "The *real* cost comes when we don't work on developing our people."

I guess my question is where did all of the mentors go? Where are the company officers? Where are the guys who have the information and experience that's needed to teach the new firefighters how to survive both in and out of the firehouse. Who can teach them the stuff they need to know? Isn't that our responsibility? You're out there, guys, and you need to share the wealth. Share your knowledge. You used to do it. We need to get these guys with the experience to give before they leave, because once they leave, they're gone. And all of that experience and knowledge is gone with them.

Before They Walk Out the Door

> "The success of an organization depends to a great degree on its leaders and their capability to supervise, inspire, and train their people."

Once our experienced members walk out the door, they're gone forever. Don't let everything that they have learned and worked and sacrificed for go to waste. Sit them down with the new guys and get them to share their knowledge and experiences with them. This is when you'll hear that complaint that "they won't listen." I know it's difficult but you need to make them listen. This is your chance.

We Used to Give Orders

We used to sit the new guy down and explain what was expected of him or her to keep things the way we want them, which is the way the guys before us wanted them. I can remember my first day on the job when my lieutenant, Bill Allen, sat me down while he explained why we take care of the rigs, our tools, our firehouse, and more importantly, each other. He explained what was expected of me, said he knew my dad was a firefighter and that I had "some" firefighting experience (I thought I knew a lot, and, boy, was I in for a big surprise!), but that those things didn't matter right now. Everything was new and my career was starting right at that moment. I knew that Bill was going to make sure I met those expectations. He explained what we were all about and that the public didn't owe us anything. He said that, in fact, we owe them for giving us the opportunity to work in the best profession in the world. He also pointed to a firefighter in the day room and said, "You see that firefighter over there? That's Dan Godfrey. He's got over 30 years on the job. If I ever see him pouring his own cup of coffee with you standing next to him, I'll have your ass. Take care of him. He's a good firefighter. It's called respect!"

Respect the Job and Each Other

Bill said to never allow anyone to be disrespectful of the job or each other. His advice was pretty simple stuff when you think about it. We do have a choice in how our new firefighters start out. Tell them why this job is the best job in the world and why we do what we do. Share with them the history, both the good and the bad. Tell them where it all started. Several years ago, at the Fire Department Instructor's Conference (FDIC) in Indianapolis, Indiana, I heard someone say, "What's with the old guy?" as they pointed on stage to Ben Franklin. Talk about missing history (fig. 3–1).

Fig. 3–1. Ben Franklin is considered by many to be the father of the fire service.

Tell new members why fire engines are red and that the color red in the fire service stands for courage and valor (fig. 3–2). Tell them where the pike poles came from, that they were whaling hooks used in Jamestown to pull down the shacks on each side of the burning one in an effort to keep the fire from communicating to the other structures. (Firefighters even practiced exposure control back then.) Tell them where the Maltese Cross came from and what it stands for (fig. 3–3). If you don't tell them, how can you expect them to take care of them? How can you expect them to understand and truly appreciate our traditions?

Fig. 3–2. For many, the color red in the fire service stands for courage and valor.

Fig. 3–3. The Maltese Cross has stood as the symbol of a firefighter for a long time and the story behind it is fascinating.

THE STORY OF
THE MALTESE CROSS

The badge of a firefighter is the Maltese cross. The Maltese cross is a symbol of protection, a badge of honor, and its story is hundreds of years old. When a courageous band of crusaders, known as the Knights of St. John, fought the Saracens for possession of the Holy Land, they were faced with a new weapon unknown to European fighters. It was a simple but horrible device of war. The Saracens' weapon was *fire*.

As the crusaders advanced on the walls of the city, they were bombarded with glass bombs containing naphtha. When they were saturated with the liquid, the Saracens threw flaming torches into the crusaders. Hundreds of knights were burned alive while others risked their lives in an effort to save their kinsmen from painful fiery deaths. Thus these men became the first firemen, and the first of a long line of firefighters. Their heroic efforts were recognized by fellow crusaders who awarded each other with a badge of honor similar to the cross firefighters wear today.

Since the Knights of St. John lived for close to four centuries on the island of Malta, in the Mediterranean Sea, the cross came to be known as the Maltese cross. The Maltese cross is your symbol of protection. It means that the firefighter that wears this cross is willing to lay down his life for you, just as the crusaders sacrificed their lives for their fellow man so many years ago. The Maltese Cross is a firefighter's badge of honor signifying that he works in courage—*a ladder rung away from death.*

It's a Privilege

Make them listen. Remind them that it's a privilege to be a firefighter. And that we owe the public a service. Help them understand that it's an honor to be part of this family and how great it really is. And if they don't like it, show them the door, because if they don't have the passion now, we'll lose them for sure down the road. They'll become the 5 percent who don't care.

Share the history

If we don't share our history with new firefighters now and they don't understand or appreciate it, what's the fire service going to be like in 20 or 30 years? We might just as well throw out all the uniforms and wear nametags that say, "Hi, my name is Chip." There are firefighters out there right now who don't have a clue where their collar insignia came from or what they stand for. They don't know where the speaking trumpets came from; they're not the kind of bugles that play music. They're speaking trumpets that were used to give orders years ago (fig. 3–4). Tell them everything about our fire service. And then go back and review the fires that have occurred in our country and discuss the impact each one has had on the fire service as it is today.

Fig. 3–4. Speaking trumpets were used for giving orders and are now used as a symbol of rank within a company or department. Those pictured stand for the rank of captain.

Time for a History Lesson

Explain to the new guys where it all started with *your* department. Tell them about your department's history. Who was there before you and what kind of impact did they have. Tell them where your department needs to go *now*. If you don't know, find out! This is our chance. Set the tempo right from the start. Pull the new firefighters into the circle now and then they can begin to appreciate how great this profession really is.

Bring back the mentors and train

Mentoring people means giving them the stuff that they need to learn and improve. Is a mentoring program really that important? Why do we need a mentoring program anyway? These two questions pretty much answer themselves (fig. 3–5). But still, from time to time you will hear someone ask, is it worth the work? Do we really get anything out of such a program? And when will we see results? The answers to these three questions are *yes, yes,* and *almost immediately.*

"Leaders must possess a real understanding of the mission of their organization."

Fig. 3–5. The more seasoned firefighters have to share their experiences and wisdom with the younger firefighter if we are to see a decline in firefighter fatalities. Following in the right footsteps can take you a long way.

Is It Worth the Work?

Yes, a mentoring program is worth the work. You could apply this question to just about everything we do. Is it worth the work needed to keep your apparatus looking good and operating properly? Yes. Is it worth the work that is needed to take care of your tools? Yes. Is it worth the work that is needed to take care of all of the little things that need to be done? Yes. And is it worth the work that is needed to know your job better, to learn a little more about what we do? Of course it is. It comes down to what kind of firefighter you want to be, and, if you're a leader, what kind of firefighters you want to work with. Note that the question was what kind of firefighters you want to *work with* as a leader, not what kind of firefighters you want *working for you*. Working *with* your members—involving them early in their career—promotes a team atmosphere and usually results in a troop that will do just about anything for you because they know that you care about them and consider them essential to the team's success.

There are a few fire departments out there (and note that I said few) where it seems to be a roll of the dice that determines whether the apparatus starts or if anything on it works. When it comes to taking care of their tools they're usually rusted, stuck to the side of the apparatus, or beat to heck. When members from these companies try to use their Halligan tool, they tend to use it like a plunger. They're the same groups you have to fight with to get them to turn off Jerry Springer and learn a little more about their job.

It's Worth the Work

Remember, anything that makes our fire service better is worth the work. A good mentoring program is a great start and can serve as a pretty good foundation. The mentoring concept can vary from department to department. A lot of it will depend upon your budget, any programs that are available to your department, the attitude among the troops, and whether or not the support is there from the boss. A lot of what we do when it comes to mentoring doesn't cost us any money, but failing to develop our people can cost us severely. The bottom line is, make sure that your experienced members are sharing their wealth of knowledge, and are making every attempt to help develop other members.

Do We Really Get Anything Out of It?

Yes, you will get something out of a mentoring program if you are willing to get something out of it. The person being mentored needs to have an open mind and be willing to listen. Likewise, the mentor needs to have an open mind, and, more importantly, be a good listener. This will provide you with a system of checks and balances that enables you to see whether what you are doing is working or not. Good mentoring doesn't mean just giving out the answers to a test. It means giving out the "stuff" that the new members need to grow and succeed, and it means giving them *all* of the stuff. All too often you'll find some people who hold back information for fear that "they'll know as much as I do." Most of the time you'll find that it's just someone feeling a bit insecure. Remind everyone that having every firefighter know as much as the next is a good thing and shouldn't be anything to fear. We should want to give them all that we can. It's the only way that we can keep getting better as a group. So give the new firefighters the things that are going to help them now and which they can build on well into the future. When we get people to know their jobs better they will do it more safely and that should be number one on the goal chart.

When Will We See Results?

You will see the results of a good mentoring program almost immediately! A good mentoring program will allow those in it to feel a sense of association, that they belong, and that someone is willing to give them what they need to do their job better and to develop. Those being mentored will recognize that there are people out there who are willing to share, to sit down with them and say, "Here, take it." On the other hand, if your mentoring program, despite your good intentions, makes it difficult to share or receive information, then the people in it get frustrated and it becomes one of those things that we don't enjoy and don't want to do. Please keep in mind that this section is not intended to serve as a step-by-step guide for developing a mentoring program, but rather is written in support of the concept. To end up with a good program, you need to determine what it is you are looking for and what you want to get out of it.

Where Do We Begin?

Where you begin depends on what you want to accomplish. Do you want something for a specific rank or are you looking for a program that addresses the whole department? Some departments work at developing a mentoring program that deals with a particular rank, such as captain. The feeling is that this is a great place to start because the captain holds a key position within our rank structure. Our next battalion chiefs come from among the captain's rank. More importantly, the captain is the individual who is going to make it all work with the guys. The captain's role is critical when it comes to development of the rank and file firefighters. Another reason that the captain may be a good place to start is that the numbers are usually easier to manage. We're starting to see a lot of captain's academies popping up across the country and many of them are good, solid programs such as those offered in Lewisville and Arlington, Texas for example. These captain's academies are covering topics such as budgets, policy development, tactics and strategy, firefighter rescue and safety, and team building, and tend to cover all the ground that is needed to be a successful captain. The students just have to be willing to take what they learned and put it to work. Those who are putting these programs together need to keep in mind that your goal shouldn't be to have the graduates go out and teach physics, but rather to go out and be good leaders and firefighters.

The Lewisville Program: Tapping into Years of Experience

In Lewisville, we started with a program that addressed the needs of the rank and file and to this day we are continuously looking for ways to improve it. Many of the suggested improvements came right from "the guys." They'll tell you what they need if you let them. We started with something called the "Mentor Questionnaire." This all came about when we realized that we were losing some of our people due to retirement and that when they left, they would be taking all of those years of experience with them. We wanted something simple that would capture the information while we still had a chance. The goal was to grab this information from our senior members, *our mentors*, before they left us. They were asked to take a few minutes and jot down some things that would tap their experience, knowledge, and tricks of the trade. The information they shared would be passed along to our new and future members.

The Five Questions

The instructions were easy. Answer the five questions, and, most importantly, keep it simple! You don't have to answer them all but it would be appreciated if you did. Answers can be typed or hand-written and don't worry about spelling or grammar. When it comes down to it, spelling and grammar haven't saved a firefighter's life yet, but the information you provide could very well be the difference between whether a firefighter goes home or not. The following five questions were asked of all members with fifteen years of experience or greater. The questionnaire is also given to a member to be filled out when that member has put in for retirement. The questions cover the following topics:

1. Apparatus, tools, and equipment. This question is asked in an attempt to identify any special needs or quirks with our apparatus, tools, and equipment. We want to know if there are any safety concerns or potential problems.

2. Calls, incidents, experiences, or problem buildings. With this question we're looking to gain information in several areas. We want the responders to tell us about any calls, incidents, or experiences that offer a "Lesson Learned" or could serve a historical purpose. The one area we like to emphasize gaining information from is problem buildings. Time and time again you hear about a bad incident or one that involves a firefighter fatality, and during the critique or investigation it is revealed that someone in the department or company knew about the building and its problems or the special danger it posed, but never shared that information (fig. 3–6). This is the kind of information we don't want our members with experience to walk away without sharing. Because when they're gone, they're gone, and those experiences go right out the door with them.

3. Lake Lewisville or dive operations. We'll take any information that they can give us about our Lake Lewisville operations. We want to hear about hazards, safety concerns, suggestions, or anything that will help a new member stay safe, especially when it comes to diving in black water.

4. Department history. Many fire departments have done a great job of preserving their department history. One in particular, Coeur d'Alene, Idaho, has done a great job in preserving their artifacts and holding on to the information that founded their department (fig 3–7). We asked this question in an effort to reclaim some of our heritage while at the same time trying to gather information about our department's growth and history so that each firefighter brought on in the future will be able to see where it all started.

5. Your wisdom and thoughts. This question is easy. We want responders to tell us anything that they would say to a new firefighter that could help him or her stay healthy and safe throughout their career.

APPARATUS, TOOLS, EQUIPMENT

Know the apparatus you are assigned to. Know it front and back. If you are sent to get something, you should be able to retrieve it by opening the correct compartment the first time. Know how to operate all the equipment on the apparatus. Know the quirks of all the power equipment. There may be three identical saws, but each one will start and run differently. Practice starting and cutting with them. You may use it to save a life someday. Somebody may use it to save YOURS! Know how to use all of the hand tools also. You know what a Halligan bar is, but can you use it properly? Can you and your partner use it to force a door? Who holds, who strikes, who calls the blows to the Halligan? The answer is in the book *Safety and Survival on the Fireground* by Vincent Dunn.

The apparatus are all different also. Even if they look alike, they will each have a personality of their own. Know what they will and will not do. Take care of them. Keep them clean and in good order. The citizens bought them for us. They deserve to see that we are taking care of something that will help take care of them.

What is my favorite apparatus? The truck, because I enjoy truck work. However all of the apparatus appeal to me, especially the reserves. They were here when I started or arrived shortly after. They have made a lot of calls.

Submitted in 2001 by
Firefighter/Paramedic Gary Apple
Date of hire: 5/1/85

Fig. 3–6. We've all had that one building in which we hoped to never fight a fire. The next step is to share it with the rest of department and with those who come aboard in the future.

CALLS, INCIDENTS, EXPERIENCES OR PROBLEM BUILDINGS

One problem building is Central Baptist Church, 407 Hickory Street. The building has the original roof with a room and enlarged roof over this roof for an extension and attic space and hidden rooms (back draft potential).

All of Old Town Lewisville, study and be familiar with its construction and outdated codes. Unique, hazardous and dangerous.

Most of your eating establishments have center core construction of a bar, with liquor and storage, with concentrated loads on roof, center of construction. Roof collapse is imminent.

Submitted in 2001 by
Captain Randy Cade
Date of hire: 2/16/76

Fig. 3–7. What are you doing to preserve your department's history? The Coeur d'Alene Fire Department in Idaho created their "wall of flame" to share and preserve their department's heritage.

DEPARTMENT HISTORY

One of the most important incidents that changed our department was the fire at Oak Tree Village Apartments, in 1983. We lost about 290 units. This fire was the turning point in Lewisville, and caused us to upgrade our tactics and consider adding aerial capabilities.

The addition of large diameter hose was another of those defining moments that changed our tactics and our philosophy. We were one of the first, if not the first to go to LDH in North Texas, and it gave us the reputation of being aggressive and innovative in the fire service. The addition of LDH allowed us to take our apparatus to the scene instead of to the plug, and additionally gave us the ability to have a "quick strike."

I think we lost a lot of tradition in our department when we sold two 1947 American la France convertible pumpers. They were classics that I wish we still had around today.

Submitted in 2001 by
Battalion Chief Jerry Cunningham
Date of hire: 2/16/76

LAKE LEWISVILLE OR DIVE OPERATIONS

Our lake swim areas have drop offs. This is where we tend to lose people, they are walking along and suddenly the wter is over their heads. Get someone to show you these areas, most of the divers on the dive team are aware of these spots.

Submitted in 2001 by
Battalion Chief Darrell Brown
Date of hire: 9/16/83

YOUR WISDOM AND THOUGHTS

I do have some thoughts that a young firefighter could possibly use. I have never thought of myself as a really wise person who should be offering advice, but a few things come to mind that others may find useful. Think about keeping a personal journal. Write down a brief summary of each shift including who you worked with, any interesting calls, important dates or special days in your career. I can't even tell you when I promoted to driver or captain. I was talking to some recruits just last shift. We were talking about a particular fire on a particular street, but I could not remember any details. This has happened so many times. If I would have needed some advice, my entire career would have been documented. I plan on starting one now... after $16\frac{1}{2}$ years.

One positive, I call it positive, about the fire service is that it requires so much of you. Let me explain. A firefighter who feels comfortable enough in their career to sit back and relax is in danger. This service requires you to exert a tremendous amount of personal energy just to keep up with the changes. Not to mention all of the younger firefighters who are getting smarter and will soon be passing you in rank. You can't relax in this field; you must stay sharp in order to enjoy a long career. If the senior firefighters that you live with seem to spend more time doing personal projects or reading the newspaper and you never see them reading fire service related articles, be careful to call them your mentors. Every shift you need to learn something new and review something you used to know. It is true; if you don't use it you will lose it.

We describe firefighters with many words. One that we seem to forget about is flexible. You will have a much better career if you are flexible. I have been in this department for only $16\frac{1}{2}$ years and I can assure you that the best words to describe this department for the past several years is always changing. If you don't like what is going on today, wait until tomorrow. Things will be different. An ever changing environment requires flexibility.

I have always asked my different crew members for about 10 minutes of 100% concentration every shift. This is the time from alarm to when we arrive and solve our customer's problems. I don't think that is asking much at all. I feel that it is so important for every member of a responding unit to do their own size-up en route to the call. Have fun in this career...it is indeed the ultimate team sport.

Submitted in 2001 by
Captain Jerry Wells
Date of hire: 5/1/85

The Mentor Book

All of this information is then taken and placed in our "Mentor Book." This book contains a table of contents, background on the questionnaire, and five sections. Each section is for one of the five questions that were asked. As you look at each section you can read the information provided and see the name of the firefighter or officer who provided it, the date they entered the fire service, and the date they submitted their answers. Also in this book is an explanation of our promotional process, the process to follow to "bump up" or "act up" at the rank above yours, and other items often found in a career handbook. Every new member is required to review it and copies of the book are kept in all stations, at the administrative offices, and at the training division. What we have noticed is that some of the more seasoned members are calling the firefighters who contributed information to ask questions about the items they submitted. We have also found that some of the more senior members weren't aware of a bad building. So this book offers benefits both up and down the experience ladder. Our goal is to have a book that is full of great information that will help keep our members healthy and alive.

"A leader must be genuinely interested in what is going on."

It Has To Be Fair to All and Objective

In other areas, our recruitment and retention committee has looked at our hiring process and made recommendations and many of those were implemented. We also looked at our promotional processes and restructured all of them. Previous promotional processes were riddled with favoritism and subject to those changes that seem to come up right in the middle of the process and leave everyone with a bad taste in their mouth. Since we implemented the new structure and process there have been no complaints and several compliments. This is due mainly to the fact that the promotion is done fairly and objectively and is now a system that allows those involved to prepare properly. And by the way, this process was put together and implemented by a committee made up of members from the rank that was being tested and some from the next level above and was not a process designed by "The Chief." We allowed for participation and with that came ownership from the guys who designed the process.

The next step in our mentoring process is what we refer to as *positional line of sight mentoring*. This is the process of training a member for their next position or promotion—you know, the one you can see but just can't reach yet. Our goal is to get everyone ready for his or her next step.

"Positional line of sight mentoring."

From Firefighter to Driver Engineer

As a firefighter gets to a point where he or she feels that they are ready to "bump up" or when their captain feels they are ready to act up, the captain submits a request to the battalion chief, who then submits it to the training division. The member is then given the material they need to study for their written exam and a time frame for completion. After completing and passing the written exam, candidates are given a driving and pumping practical. Both require a passing grade and candidates are given plenty of time to practice. The idea here is to get them ready for the promotional process for driver engineer where they are given a written exam along with a pumping and driving practical. The driving practical is the same one given during the promotional process and the pumping practicals are similar. When all is completed, the training division submits the member for approval by the chief of operations to serve in the acting driver engineer role.

From Driver Engineer to Captain

This process, too, starts with a request or recommendation to act up as captain. The promotional process for captain involves a written exam, a tactical assessment, and an in-basket exercise. The process for a member to be approved as an acting captain is pretty much the same. They are given a written exam, a tactical assessment involving a single family dwelling fire, and several in-basket exercises dealing with the types of scenarios a shift captain would face. After candidates pass all portions of the exam, the results are sent to the chief of operations for approval.

From Captain to Battalion Chief

This process is similar to the one described for a driver engineer looking to act up as captain except that it is tailored for the multi-company officer. The tactical assessment involves a multi-family dwelling fire with several challenges, and the in-basket exercises hit on the types of situations that a battalion chief would handle. In addition to the above requirements, candidates have to ride out with the battalion chief for a minimum of three shifts. During those shifts the battalion chief and the captain can switch roles, allowing the captain to act up as battalion chief and the battalion chief to serve as a coach.

Division and Assistant Chiefs

Our division and assistant chiefs often work with each other to prepare for the next level. The idea is not to hold anything back. As I stated earlier, too many officers are afraid to share what they know with those coming up under them because "then they'll know as much as I do." But sharing information is the whole idea and is the only way to make sure the fire service gets better.

Mentoring shouldn't be just a buzzword and it shouldn't be something that is hurried together just so you can say you have a mentoring program. At the same time, it shouldn't take years to develop or get through. You can always change or enhance your program. Think back to the times when one of your mentors sat down with you and shared. Think how valuable that information was and probably still is. Think about all of the "stuff" that our mentors have given us. A good mentoring program has to touch on several levels and cover a variety of topics. The overall goal should be to spread information up and down the experience ladder. Give them what they need to do their jobs and help prepare them for their future.

It is mind boggling to consider what we can do with the information that we can grab from our senior members before they leave and from those who will be around for a while yet. A good mentoring program will make your incidents go a lot smoother, elevate your organization overall, and most importantly, provide a much safer environment for the troops to work in. We owe it to our members to have good mentoring programs.

Define Expectations

So start a mentoring program and take the time to let your firefighters know what the expectations are of them. It's a little hard to get on a guy later when you didn't give him the game plan up front. We're killing over a hundred firefighters each year and injuring tens of thousands in spite of having better apparatus, better gear, and better tools. There's no new fire out there killing us. It's the same stuff that's been killing us for years.

Firefighter Survival Training and Rapid Intervention Teams

While we're on the subject of losing firefighters, I want to encourage you to teach your members firefighter survival training. Teach them how to get themselves out of trouble, survive, and go back home (fig. 3–8). Don't let anyone bully you into thinking you'll hurt your guys teaching them how to survive. Train safe, but train. You know, I can hurt myself with a sledgehammer if I try hard enough (fig. 3–9). We train our people to work with saws that travel at 6000 revolutions a minute (or 250 miles per hour), to work with hydraulic tools that will lift a school bus off of the ground, and to crawl into burning buildings. It's a dangerous job and when you freelance on the fireground you risk killing or injuring your firefighters. When you freelance on the training ground you risk doing the same thing. Train safely and as if your life depends on it, because it does (fig. 3–10). To the guys who have the brass axes to train their people in firefighter survival techniques, and especially those of you who have little or no support from your administration: keep going. Keep fighting and teaching your guys how to go home even when things go wrong (fig. 3–11). About the only thing you're going to do wrong is save a firefighter's life. And one more thing, RIT works! Just understand how many firefighters it takes to make the rescue (fig. 3–12).

Fig. 3–8. Training in firefighter rescue is extremely important. This particular rescue drill (the Denver Rescue) was created in a response to the tragic loss of engineer Mark Langvardt who died in the line of duty in Denver, Colorado in 1992.

Fig. 3–9. Chief Tom Shervino demonstrates a portion of the floor drag drill at FDIC in Indianapolis, Indiana.

Fig. 3–10. The second floor rescue of a firefighter is a very labor intensive task but one worth practicing.

Fig. 3–11. Probably one of the most difficult rescues, getting a brother up the stairs for a stair rescue has challenged firefighters many times.

Fig. 3–12. RIT does work! Just remember there's a lot to it and that rescuers have to have the right attitude, just as the RIT team in the picture does.

Fight for new equipment and be honest with what's killing us

Remember one thing when you're crawling down that dark, hot, snotty hallway: the air pack on your back, the nozzle in your hand, and the protective clothing you're wearing, were all bought from the lowest bidder. How does that make you feel? When you look at the reasons for firefighter deaths and examine the contributing factors, the same things keep reappearing: lack of command and control, lack of an accountability system, poor communications, failure to follow SOPs, failure to read the building and the fire properly, and a long list of reasons showing us that we need to get back to basics. The words of one of my mentors are worth repeating here: "You want to be a good firefighter, you need to know building construction and fire behavior. You have to know how the building is going to react to the fire and how the fire is going to react to the building. You need this before the rest."

Fireground—Battleground

On that same note, the late Frank Brannigan (keep in mind that referring to Frank as an expert in building construction is an understatement) has said for a long time that to compare the "battleground" to the "fireground" is a very true comparison. But, Frank says, we differ from military strategists in the following way. They are often successful in their battles because they study their enemy. They learn everything about them. They know what their strengths and weaknesses are and they use that information in their strategy.

> "Remember one thing. When you're crawling down that dark, hot snotty hallway: the air pack on your back, the nozzle in your hand, and the protective clothing you're wearing, all was bought at the lowest bid."

> "You want to be a good firefighter, you need to know building construction and fire behavior. You have to know how the building is going to react to the fire and how the fire is going to react to the building."

We don't. He reminds us that our enemy is "the building" and that we just don't spend enough time studying our enemy, the building. Today, a firefighter who doesn't study building construction is weak to say the least. There are good classes being taught by good instructors and there are great books covering building construction as well. Don't sit back and wait or guess what a building is going to do or how it's going to react to fire. Go out and become a student of building construction. I promise you, it will make you a better, smarter, and, most importantly, safer firefighter.

Our leadership

As you continue to look, you see that our leadership or lack of leadership, often allowed for some of those bad things to happen. Often the leadership problem starts back in the firehouse. If they're a bunch of mutts in the firehouse, they're going to be a bunch of mutts on the fireground. And mutts are not always just chiefs. You can find mutts at every level. There's no metamorphosis that occurs on the fireground. The attitudes start in the firehouse and carry straight through to the fireground. Another good friend of mine, FDNY Battalion Chief Don Hayde said, "Don't blame the guys in the company. Blame the company officer. He's the one that allows it to happen. Go after him or her. Hold them accountable." And it's true. If we can't trust our leadership in the firehouse we sure can't trust them on the fireground. I mentioned earlier that the company officer sets the tempo and attitude for the firehouse and can affect or influence that of the entire shift (fig. 3–13).

We can all remember the good officers we have worked for and how they have impacted our lives. They made a difference. They were there for you when you needed them, they didn't leave you in times of crisis, and they understood you. At the firehouse we refer to them as working officers. They don't mind getting dirty once in a while, they help you with projects, and they are really part of your crew. They don't hole up in their office and keep distant from the crew. They are fair and honest and don't pull any punches. They care enough to spend the time and effort with you and want to see you succeed, but most of all they want you to go home at the end of the day. We're losing too many very special people each year not to care. Not to try. The bottom line is that we do have a say. We can provide a foundation for pride and stimulate the attitude that is needed. Remember, the one thing in life you have absolute control over is your attitude.

Fig. 3–13. Chief Don Hayde instructing firefighters in roof ventilation.

Your circle of influence

The extent to which your attitude and actions affect the people you come into contact with is called your circle of influence. Used properly, it can influence people to do the right things. If you lead by example and "walk the talk," if you will, you can pull people into your circle of influence. Be positive, say good things, be nice to your people, and watch your circle of influence grow. The fence walkers will eventually fall in and the 5 percenters will go away. It can also work the other way with a bad attitude. When that happens, you have to take immediate action and eliminate the source of the problem. The company officer can have a positive effect in the firehouse or on their shift, or could be poisoning the water. The opposite of the positive company officer and role model is the one who sits at the kitchen table and holds court, tearing anything positive to shreds and destroying morale. The problem is that this kind of officer usually blames someone else for the morale problem, not realizing that they are one of the contributing factors. If you're trying to figure out whether you have a positive or a negative influence, ask yourself the following questions:

- Do I start the rumors or stop them?

- Do I exhibit a positive attitude or a bad attitude?

- Do the troops hang around with me because I let them get away with murder or do they hang around with me because they know I care and will protect them?

- Do I serve as an errand boy for the chief or do I stand behind his decisions and explain that it's for a good reason and needs to be done?

- Am I part of the problem or part of the solution?

- And probably the most important one of all. Am I their leader or their buddy? You can be both, but you need to be a leader first. You don't want them to follow you just because you're their bud, you want them to follow you because you'll keep them out of trouble (both on the fireground and in the firehouse). You care and you will protect them because you're a good officer.

> What you see here,
>
> what you say here,
>
> what you hear here,
>
> when you leave here,
>
> let it stay here.

Insist That They Appreciate This Job

In the fire service we come up with solutions and get rid of problems. Make it so in the firehouse and the people who cause problems will end up on the outside looking in, saying, in the words of the great Bob Uecker, "Boy, it sure looks like they're having a good time in there." Start on each candidate's first day and continue to build the foundation for them. *Insist that they appreciate this job.* We owe it to those who have worked so hard and sacrificed before us. Don't let the old timers come back and say, "What in the world did you do to my department?"

Learn to market your fire department

Go out and show it off. Whether you're a career or volunteer firefighter, get out of the station. Or better yet, open up the station to the public (fig 3–14). If your neighborhood will allow it, open the overhead doors once in a while and let the neighborhood see that the station is occupied. Let the guys sit in front of the firehouse in the evening. We put park benches in front of our firehouses. I want the guys sitting out in front,

Fig. 3–14. Show off your firehouse. Open the bay doors if you can and let the neighborhood know you're home and open for business.

connecting with the families in the neighborhood. Let the kids see the fire trucks. Try to make the firehouse look like a firehouse again. This one is probably going to be hard to believe, but in Lewisville we actually had a "no loitering" rule here that stated that "personnel shall not congregate, loiter, or otherwise meet in the rear or front of the station." Does that sound as ridiculous and just plain goofy to anyone else as it does to me? As long as they are not doing anything wrong, what's the big deal?

A neighboring training chief said that I ruined what someone else spent 16 years working so hard for by letting the guys sit in front of the firehouse in the evenings (fig. 3–15). I don't know how you can ruin something that ridiculous. I'll be happy to explain to anyone why we do it!

Fig. 3–15. Lewisville put park benches in front of all of their firehouses and encourages their firefighters to sit out front where they can connect with the public.

Get the Guys Out of the Firehouse

Get the guys out reading to the kids in the schools. Go out and find ways to market your fire department. See what others are doing and how it works. Go out and brag about your department and tell people why it's great. I walked into our Central Firehouse in Lewisville on my first day and didn't know whether to open up a checking account or buy insurance. All the desks and tables had glass on them. Who puts glass on furniture in a firehouse? Then I looked at the walls and said, "Where's all the fire stuff? Someone stole all the fire stuff off of the walls. They're all bare." We took care of that. There's "stuff" on the walls now! Only if it smells like a firehouse, tastes like a firehouse, and looks like a firehouse, it is a firehouse. There are a few that slip through, looking good on the outside but pretty empty on the inside, if you know what I'm saying.

Big hat, no cattle

In Texas they refer to that as having "a big hat with no cattle." But we're the *fire department*. Let's start acting like it. To those who say we have to be more like a "business" these days: you're right to an extent. But you can still look like firefighters, your firehouse can still look like a firehouse, and your firefighters can still act like firefighters and be professional.

The Neighborhood Wants a "Firehouse" Down the Block

To be honest, the neighborhood, the residents, and the business owners all want a firehouse down at the end of the street. They don't want us to be a Fortune 500 company or "business." They want a firehouse with real firefighters in it, like the ones they grew up seeing. They want a firehouse that keeps the doors open once in a while so that they see the firefighters. They want a place kids can go to for a visit or go to when they're in trouble. They want a neighborhood firehouse they feel connected to. I promise you, that's what they're looking for.

The best job in the world

This is the best job in the world. If you don't like it, get out! Send those who don't care about this job, who don't care about their brother or sister firefighters, who don't love it, a strong message. Tell them to go down the street and work for K-Mart stocking shelves. Then they can have a job that they don't have to think about when they go home.

I'm not much of a poet, but this one fits very well and I've used it in the past many times. I got it from my friend Bill Farnum. It applies not just to the company officer, but to all of us.

> *"I saw some men in my home town,*
> *I saw some men tearing a building down.*
> *With a heave and a ho and a mighty yell,*
> *I saw a beam swing and a sidewall fell.*
> *I asked their foreman are these men skilled,*
> *The kind that you'd hire if you wanted to build.*
> *He laughed and said why no indeed,*
> *For common labor is all I need.*
> *For with common labor in a day or two,*
> *I can tear down what took a builder 20 years to do.*
> *I asked myself as I walk away,*
> *Which of these roles am I going to play?"*

The message is, you can go out and be the best company officer you can be. Study the position. Talk to those who are successful. Ask them how they make it all work. Evaluate your performance constantly and always try to improve yourself. Share your knowledge and experience. Be a brother. Love this job and be passionate about it. Share information and continue to build this great profession of ours. Or, you can go out and tear it down. The choice is yours. If you do care and truly love it, keep working at it. Keep pushing forward. Go out and share our profession with somebody. Read something about this job everyday. You owe it to yourself and to both of your families. If nothing else, do this one small thing: make a difference by leaving the fire service a little bit better for the next guy.

4
THE CHIEF

From as far back as I can remember I've wanted to be a firefighter just like my dad (fig. 4–1). As I got older my feelings didn't change, even while I was a cop waiting to be hired as a firefighter. I guess I've just always wanted to make a difference. When I finally made it into the firehouse on a full-time basis, I was like a lot of young firefighters, thinking about—dreaming a little—about where I wanted to go in the fire service, this great profession. I constantly asked myself just how far I wanted to go and just how far I wanted to go to make a difference. As I moved forward, I realized that every time I moved up there was more responsibility and more to do. But it also hit me that with each step forward I got to make more of a difference and have more of an impact, mostly in a good way, though not always. Being human is also making mistakes and I've made my share.

Fig. 4–1. Your level of responsibility increases with every promotion

The Big Chair

As I traveled along, I decided that if I ever got to sit in the "big" chair I would do all that I could to avoid being the kind of chief that I hated working for. Early on, I asked a friend of mine how to do that and he said it was easy. He told me to take a look at the guy that no one wants to work for, and do the opposite of what I saw him doing. So far it has worked! But I was blessed early in my career to have worked for several good fire chiefs who were also really good "bosses," like Dick Vachata, Ron Szarzynski, and Bob Rubel.

Chief Vachata was easy to figure out. Love the job and take care of each other and you were gold in his book. Chief Szarzynski had a similar way of thinking. If you were good and did your job, he was there for you. You knew what lines not to cross, but he was there for you all the time. Chief Rubel was a flat-out good chief who didn't mind giving you a kick in your "sector C" when you were screwing up. Hey, I have no problem admitting that he saved my career with some very strong advice when I was going through a very bad time in my life. I will be indebted to him forever. If it wasn't for him, I don't know what I'd be doing right now. Years later, I wrote him a letter to thank him for what he did for me.

My goal when I became chief was to never lose focus or lose touch with what the fire service is all about. I went to a chief's conference a few years ago and was startled to see and hear from so many who had lost their grip on what it's all about. I came away from there hating chiefs more than ever. Many of the chiefs I ran into could only talk about how rotten their guys were and how they wished they could get rid of them and privatize, and that union, and blah, blah, blah. I thought to myself, "My God, where did these guys come from?" I just had to look a little harder because there were a lot of chiefs at the conference (and there are a lot of chiefs nationally) who do care, and do realize how important their people are, and do support their people. They understand. I truly believe that there are a lot more chiefs on the good side than there are on the "dark side." It's just like anything else; the negative ones tend to stand out.

My favorite saying: egos eat brains

It seems like everything rolls along just fine until someone's ego runs amuck. I have seen several chiefs who started out with good intentions, and then after a few years, allowed their egos to grow out of control and to a size that prohibited good things from happening. You may have seen a situation in which someone let their ego get out of control and it ended up costing them relationships, friends, and in some cases their job. I know one individual who let his ego get so far out of whack that the city administration didn't trust him or listen to him. He could have offered them the cure for cancer and they still would not have listened to him. The best choice is to keep yourself humble, to poke fun at yourself once in a while, and to stop taking yourself so seriously. It isn't about you! Sometimes you're just not as smart as you think you are.

> *"Stop taking yourself so seriously."*

Attitude

"The only thing in life that you have absolute control over is your *attitude*." It is your choice. You can choose what kind of attitude you're going to have today. It's true. Anger, happiness, sadness, and so on are all emotions, but it's your attitude that can make you or break you. They can steal your car, say terrible things about you, or find other ways to hurt or anger you, but it's your attitude that can provide you with the right path to take. Remember, decisions made on emotions have hurt and killed people. It's the decisions that are made on good solid ground and common sense that go far. Your attitude is about the only thing that they can't take away from you or at least without a good fight!

Earlier, we discussed your circle of influence and how it can affect those around you. Remember that your influence can go both ways: good and bad. When your attitude is on the downward slope it can be contagious and rub off on those around you, especially the newer guys. But when your attitude is positive, there's no telling how far it will take you and the people around you. Your circle of influence can have a tremendous impact on your organization and, more importantly, on your own life.

> *"The only thing in life you have absolute control over is your attitude."*

Remember to place some emphasis on your values. We also discussed the idea that, with solid core values and reasonable, realistic vision and mission statements, you can create a foundation for success. There are departments that sent someone to a mission statement-writing class, who came back and wrote a ten-page-long mission statement and then tried to fit it on a coffee mug or the back of a business card (fig. 4–2). They were enthusiastic and had good intentions, but they lost sight of what a mission statement is supposed to do. They put something together that is so big and so wordy that no one can remember it all. And if you can't remember it, how can you support it and believe in it? It may look good on paper or on the cover page of a strategic plan, but it won't work as it was intended. Your values or guiding principles need to serve as the foundation for it all. They're what your mission statement is built on, and if done right, is the only way to realize your vision.

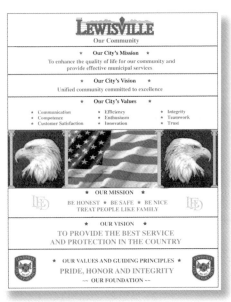

Fig. 4–2. Lewisville's mission statement. Short, simple, to the point, and it supports the City's overall mission and vision

Lacking Foundation

So many people and organizations struggle because they lack a solid foundation of values. Today, we're facing some pretty serious challenges. Families are struggling over values and our country's foundation is being stressed and cracked. Many organizations forget to establish core values, or if they have them, forget to follow them. It's hard to read a newspaper or listen to the news without learning about a corporate executive being investigated or indicted for stealing from their own firm, or worse yet, from their employees, their corporate "family." It's hard to work within or support an organization that doesn't have integrity at the top. It's hard to support a thief. All of this is why our communities need the fire service now more than ever.

It all comes together pretty easy if you remember to treat people like family (fig. 4–3). There are so many stories about people compromising their integrity or treating people like garbage. At times the stories seem so unbelievable but then you find out that they are true.

Fig. 4–3. Visitors entering the firehouse see the "our kids" bulletin board and soon after realize that if our kids are important, so are theirs

Accountability

A lot of it seems to fall right on that accountability thing and I'm not talking about a "passport" or a "PAR," but for those you are accountable for. You're accountable to your boss, your guys, the public, and your family. You're accountable to your boss to do what your supposed to do—your job. You're accountable to your guys to train and look out for them, and ultimately, to their families. Before you let them do something you'll regret later, think about how you'd explain it to their family if they got hurt. Most important, you are accountable to and for your family. Just as you would tell someone that getting into an accident on the way to a call doesn't help the victim at all, remind yourself that when you don't take care of yourself both emotionally and physically or take the steps necessary to be safe, you're not doing your own family any good. Think about your decisions and who and what they affect. What will it take to make you realize that you need to make decisions that will protect your family and your people first, before you decide what's best (or the most fun) for you.

Remember where you came from, Chief!

It really helps to have a leader who knows what it's like to work shifts, to sleep in a firehouse, or, as a volunteer, to get up in the middle of the night knowing that you have to go to your regular job soon, and who can find the fireground without the use of a compass and a flashlight.

"Managers enforce rules; leaders promote values."

Where did it all start for you, boss? Do you remember when? Or is it kind of a cloudy memory? Maybe if a couple of those wearing "5" on their collars remembered where it all started we wouldn't have ridiculous rules like "No loitering or congregating in front or rear of the firehouse." Maybe we would have decent, comfortable furniture to sit on instead of sandbags and we would no longer have rules that tell you when to turn the television on and off, and (one of my favorites) "You can't go to the grocery store while on duty." What's funny and somewhat amazing is that the guys who are coming out with these rules are the same guys who would have been screaming about them years ago when they were on shift! I'm not saying to give away the store; just try to be a little more understanding and flexible.

Good Communication = Good Labor Relations

Maybe if we all worked a little harder on communication, labor relations would work better and we could all work together. There are some departments doing really well in this area already. There are good folks on both sides of the table. Everyone just needs to remember where we all started. Help educate and inform our newer members what it's like in the upper ranks, and if you're in the upper ranks, remember what it was like when you started out. Do whatever you can to avoid becoming a "Desk Commander."

Learn about people

Years ago one of my goals was to teach for the University of Illinois Fire Service Institute. There were few field staff positions and getting in the door was tough. I had waited for a few years and had just about given up when Chief Jack MacCastland called me and asked "How do you like the Instructor I program?" I told Mac that it was all right but that I really wanted to get into the Tactics and Strategy program, which covers live burns and the "fun" stuff. Mac reminded me that this was an opportunity to get in the door and start teaching and told me not to blow it. He also reminded me of a story about John Hojek, an FDIC instructor and great chief officer. When Johnny was offered an opportunity to teach at the institute in the Fire Prevention program, Johnny said, "Fire prevention? I don't want to teach fire prevention, are you kidding me?" I laugh when I think about that story, but it was John's way in. Think about how many firefighters' lives would have been affected if they hadn't had the opportunity to learn from Johnny, one of the best fire service instructors in the country.

People Size-up

It was teaching the Instructor I through IV series that had the greatest impact on my career and my life when it came to understanding what made people tick. When you have to teach people how to teach to others, how adults learn differently than children, and what motivates people (thanks in part to Abraham Maslow and his motivational pyramid), and when you have to really look at what's going on in people's lives and why they do the things they do, you become pretty good at "people size-up."

Read smoke—read people

We talk all the time about reading the smoke and what it can tell you: what is burning, how much is burning, where the fire is going, our progress with the attack, and so on. We know how tremendously important that is. Maybe we should be reading people the same way. Just as we look to the smoke to see what the fire is talking to us and telling us about itself, so should we look at people to see what they are telling us about themselves. Take some time to look closely at people and try to understand where they're coming from. If my wife gets a little frustrated when we have a bad waitress, I remind her that maybe the waitress has had a bad day, has problems at home, works for a bad boss, or has a couple of tables with some nasty people. She's probably just trying to put food on her family's table and pay the bills. (My wife usually agrees and I really feel good because I'm usually wrong and she's right. I tell people all the time that she's the smart one in the family and thank God that our kids take after her.) Maybe if we took a little time to understand what's going on in people's lives, their reactions and behaviors might make a little more sense. If nothing else, we would definitely end up with a lot more patience. Read smoke, read people.

> *"A leader with great passion and few skills always outperforms a leader with great skills and no passion."*

Trust your people

Trust takes some work but is well worth the effort. Trusting your people goes a long way toward fostering a positive attitude. It's when you don't trust your people or feel you have to micro-manage them that they lose faith. Train them, give them what they need to do their jobs, and let them go.

A good avenue for showing trust is through performance evaluations. Let your people know they're doing a good job. Some officers leave the comments section blank or write "Does a good job" or "Okay" under each category. But if you sit down at the kitchen table with one of those same officers and ask about their people they can go on for an hour about how great they are; you would think that they could put a couple of sentences together.

The Little "Gold" Book

I like using the "Little Gold Book" principle. Every time you see a firefighter or company officer helping out another firefighter, doing a special project without being asked, or just doing a good job, make a note of it with the date and time. I'm not talking about a little black book. I'm talking about a little *gold* book. Write these things down, and when you find yourself with an opportunity, mention that you appreciate the time that he or she spent helping another firefighter study for an exam or thank him or her for helping with a problem or project. Thank your people for stepping up to the plate and doing that little bit extra, or for taking care of something that they weren't asked or assigned to do. I know that they're supposed to do those kinds of things, but once in a while they need to be thanked or patted on the back for doing them. It's as simple as saying, "Hey Mike, I just wanted to thank you for helping Larry with studying for his promotional test. You know, about four months ago when you sat in the kitchen coaching him and helping him." Or, "I wanted to thank you for taking the time to clean all of the compartments and painting the tools on the spare engine last month."

Surprise them. Don't wait for evaluation time. Pick a few times during the year and take a minute to say thanks. They will be surprised, they'll like it (if you're sincere and you don't wear them out and do it too much), and they'll be taken aback by the fact that you do notice the small things that they do. As I said earlier, catch them doing something right. Anyone can catch them doing something wrong. Oh, and the reason I call it the "Little Gold Book," is not because it's your little golden book of good little people. It's because it's like having gold in your hands. The payback is absolutely incredible!

Put Some Good Stuff in Their Files

Nominate your people for an award when they deserve one. Even if they don't receive one they'll have the award nomination in their personnel file. When it's time for an "at-a-boy," give them one. Put it in writing—maybe by using a Record of Exceptional Performance form—and stick it in their file. Post the thank you notes and letters your firefighters receive and place copies in their files. Why do personnel files have to have just the negative or "official" stuff in there? Put some good stuff in. When it's promotion time or someone is being considered for an appointment, wouldn't it be nice to see that kind of stuff in the file when you're reviewing it? Trust your people and show them that you have confidence in them. You will be surprised how far the troops will take you and the department. They're pretty smart!

Let go of the past—focus on the future

Sometimes we get so bogged down with things that happened in the past that we can't even begin to think about the future. Sometimes you need to let go of old grudges, hard feelings, and past bad times. Learn from the past; don't live in it. How many times have you seen two people who are not getting along because of something that happened 20 years ago? Or because of something that would seem silly, petty, and insignificant when it's explained to someone who isn't involved.

When I was a cop (I know; don't hold it against me!), we were sometimes called to respond to a domestic disturbance. When we got things settled down, I would look at the husband and wife and ask them if they really understood what they were arguing about. I'd ask whether the "big issue" that had brought the police into their home and over which they were ready to go to jail was that big of a deal when they really looked at it. We need to look at some of the things that bother us and ask ourselves the same question. Is it or was it that big of a deal? Hey, it's worth a try! The job is great already. Can you imagine how great it would be if when you came into work you didn't have to worry about what the other shifts were saying about you? Say nice things!

> "Learn from the past; don't live in it."

Build tomorrow's leaders and successors

Build for and set the stage for success and the future. Do you have a succession plan in place? Are you preparing your future leaders? To be honest, there just aren't enough good leadership programs out there. We need more. Go around and ask people what kind of training they got before being promoted, or for that matter, what kind of training they got *after* being promoted.

Try starting a mentoring program, sharing knowledge and experiences to develop your people. Are you getting everyone at one rank or level ready to promote to the next one? Share your successes and your failures so they can learn. Forget "the school of hard knocks" thing. Be sure that you are learning from your successes as well as from your failures. Sometimes we forget to look at why we were successful and what got us there, at where we need to improve, and where we are going next. If you don't do this, you may end up turning a good thing into a bad thing by just letting it sit there, unchanged, until we end up saying that it isn't working anymore or that we need to change directions. Sometimes, just by looking at how we succeed, we can learn something that allows us to take it to the next level, keep things on track, and keep raising the bar.

Think about how far we could go if we just laid some groundwork for the next guy. It's usually a bad sign when your department has to go outside for every position. Our city administration went outside here for a chief and they hired me and boy have they ever learned their lesson with that! All kidding aside, I work in a great place with a great group of people, but I have told my boss that I should be the last fire chief they have to hire from the outside. That is, if I'm doing my job right. Work towards the future and build for 20 years from now.

The chief's aide: a nearly extinct species making a comeback

The first battalion chief's aides made their appearance in the fire service a long time ago. The value of this position, both administratively and on the fireground, was realized very early (fig. 4–4). Probably the biggest value, though, was providing the battalion chief with another set of eyes and ears at an incident, thereby allowing the chief to make more informed decisions regarding the actions his companies would need to take at a fire. The chief's aide added plenty of value for day-to-day operations, but the impact the position had on the fireground was tremendous.

Fig. 4–4. The battalion chiefs in most cities rely greatly on their aide both in the firehouse and on the fireground. Pictured is FDNY Battalion Chief John Salka with his aide, firefighter Jack Schiavone, at an "all-hands" fire in the Bronx.

When Did We Lose Them?

Many of the large city departments had battalion chief's aides in place, as did some of the departments that were near those departments. But somewhere along the way, some departments forgot just how important the role of the chief's aide was. That fact, coupled with the never-ending budget battles that so many departments go through, began to make the chief's aide position expendable, one that was easy for some chiefs to eliminate from the budget. This was especially surprising because some of the chiefs who allowed the position to be eliminated would have fought tooth and nail to keep their aides back in their day when they used them. However, there are still many departments whose chiefs are fighting to keep their aides in place.

When you look at efforts to reduce the number of fireground line of duty deaths being made nationwide, you see the following items at the top of the list of contributing factors:

- Failure to implement and follow standard operating procedures
- Failure to use an accountability system
- Poor communications

One thought that stands out as you look at the list of contributing factors is that many can be traced back to a failure to read the building and the fire properly. There are several ways in which an aide can assist an incident commander (IC) in his decision-making role (fig. 4–5) and one way is to be that other set of eyes and ears for the IC that we mentioned earlier. Other ways, including some of the following, just make it easier, safer, and more organized to work on the fireground.

- The aide can provide the IC with another view of the building for size-up purposes.
- The aide can serve as the accountability officer controlling your incident accountability system.
- The aide can assist with communications or serve as the communications officer, leaving the IC free to think and complete other tasks (fig. 4–6).

- The aide can serve as the research officer for the IC, providing critical information on everything from water mains, to pre-plan information, to hazardous materials, and the list goes on and on.
- The aide can track apparatus in staging and/or those committed to the scene.

The chief's aide can do pretty much anything that will make the incident run more smoothly, and, more importantly, safely.

Fig. 4–5. The chief's aide can be that second set of ears and eyes at a fire and can provide critical information to the battalion chief to be used in his decision making process.

Fig. 4–6. The aide can serve as the communications officer at an incident leaving the battalion chief free to think more clearly and to handle other tasks.

Bringing a Species Back

It's been exciting to see the resurgence of this position and to see more and more fire departments asking to add this position to their ranks for the first time or to bring it back after years of absence. Considering the amount of work and responsibilities that the current battalion chief has to handle and that the rate at which that list is growing each year, chief's aide is a position that can be justified easily.

Opinions are like, well, you know the saying: everybody has one. It's our department's opinion that the chief's aide position is extremely valuable and will have a tremendous impact on our firefighters' safety. Administratively, the aide's position provides support to the battalion chief in the following ways (and these are just a few):

- Assists with the daily staffing issues

- Prepares and works through payroll issues

- Along with the above daily staffing issues including apparatus and station assignments, they work with vacation and sick call situations as well

- Injury reports

- Acts as a point of contact for the battalion chief when he or she is unavailable

The aide is a partner to the chief and part of a team that makes day-to-day shift operations run smoothly. A battalion chief who has an aide can concentrate on working with his shift instead of being tied to all of the paperwork that comes with the position.

What's in a Name?

Over the years, those trying to eliminate the chief's aide have argued that an aide is nothing more than a "chauffeur" (not to be confused with the valued chauffeur's position in the FDNY) or glorified driver for the chief. They couldn't be more wrong! Driving is just a small part of the aide's responsibilities. But while we're on the topic of driving, having someone else drive to the scene allows the battalion chief to check pre-plans, and to actually read the mobile data terminal (MDT) or mobile computer terminal (MCT) before arriving. The aide can get the chief in close, then move the vehicle or position it so that it works to the chief's advantage. By driving, an aide provides the chief with much the same opportunity that the driver of the engine, truck, or squad provides the company officer while en route to an incident. So if calling them a "driver" doesn't sit well with someone, call them one of the following or make up your own title:

- Field incident technician

- Command technician

- Chief's aide

Some More Benefits

Besides the obvious benefits related to fireground safety and administration, there is also the mentoring aspect of the chief's aide position. Some departments have assigned the chief's aide position the rank of captain because battalion chiefs are generally promoted from the rank of captain. The chief's aides are usually those at the top of the battalion chief promotional list; the aide position offers them the best advantage and learning opportunity available, one in which they are working side by side with the battalion chief. This is a program that is used currently in the FDNY. In Lewisville we were granted the chief's aide position at the rank of driver engineer. Our ranking system runs from firefighter to driver engineer, from driver engineer to captain, and from captain to battalion chief. The way we're set up the chief's aide position still serves as an excellent mentoring system for our department. Through this process we are able to accomplish the following:

- Assign one driver engineer to this position on each shift for sake of consistency

- We allow the battalion chief to substitute other driver engineers into this position to provide the experience to more of our personnel. Often this is done when the FIT (field incident technician) is on vacation, sick, or in school.

- The battalion chief can assign a driver engineer nearing a promotion to captain to the FIT position so that he or she can share their philosophies and expectations with the new captain, setting the tempo for a good transition to that new position.

- A captain being promoted to battalion chief can be assigned to the FIT position in an effort to get him or her ready to take on the role of battalion chief. This is similar to the mentoring process used within the FDNY: let the new battalion chief work side by aide with the current chief, making decisions, and handling tasks under the watchful eye of the battalion chief. This is definitely an invaluable tool in preparing for that promotion.

The list of things that we accomplish with this position really does go on and on. Another that comes to mind is related to our medical helicopter service and landing zone requirements (fig. 4–7). We land helicopters regularly to assist in transporting trauma victims to the larger trauma centers in Dallas or Fort Worth, Texas. In the past, this would require the battalion chief to respond to the scene and handle landing zone (LZ) set-up and safety, and to communicate with the inbound helicopter or helicopters (fig. 4–8). Now, the FIT is able to handle the LZ. When the incident involves a rescue, the battalion chief can be dropped off at the scene to handle incident command duties, freeing up the company officer to oversee the rescue operations.

Fig. 4–7. The aide can set up and handle the landing zone for an inbound medical helicopter, leaving the battalion chief to handle the emergency scene.

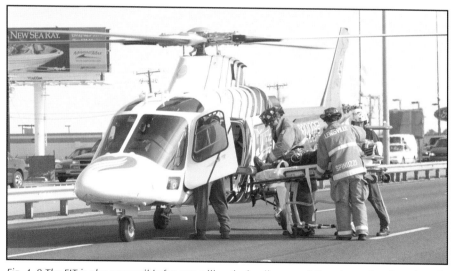

Fig. 4–8 The FIT is also responsible for controlling the landing zone (LZ) and assuring the safety of personnel operating in the zone.

Since Implementation

Since the implementation of this position in Lewisville, there has been a dramatic improvement in our department's operations, both administratively and on the fireground. Our battalion chiefs had no idea what the impact would be, but none realized that it would have as much of an impact as it has had on day-to-day operations.

The Information and Format Is Out There

We "borrowed" as much information as we could find on the FIT position before implementing it so that we would get the most out of it. We even went so far as to send our FITs to ride out with the 3rd Battalion in Dallas and with Battalion Chief Stu Grant and his aide in an effort to capture as much as we could. Former Dallas Fire Chief Steve Abraira, a pretty progressive chief, was more than accommodating with our request and helped us out a great deal. Shortly after that, the FITs were sent to do the same thing with the FDNY and a couple of their battalion chiefs and aides (fig. 4–9). To follow that one up, about a year later we sent each battalion chief with his FIT back to ride with the FDNY so that the battalion chief could work with the chief and the FIT with the chief's aide. In all three of these instances the battalion chiefs and FITs were on "our clock" and it was viewed as a training experience and funded as the same. This part of the process proved to be invaluable, providing an experience for our personnel that was second to none.

Fig. 4–9. Three new chief's aides riding out in a mentoring style process with the FDNY. All three considered it to be an invaluable learning experience.

We knew we were on track when the FIT at a multi-family fire asked what our apparatus and alarm status was, paused and said "Fourth, chief." We told him "Good job." We had been training that in a fast-moving incident the FITs needed to keep the companies coming as long as we were requesting them. That was the first time an engineer pulled a 4th alarm and it was great!

They Get Promoted Fast

I realize that lately there are a lot of departments fighting to survive: to keep their budgets from getting slashed to pieces, to keep firehouses open, and to keep firefighters from being laid off. But, if you ever have the opportunity to request the addition of a chief's aide position, go for it. I promise you, you won't be sorry. The only downside for the battalion chief—if you can call it one—is that as soon as you get a FIT trained, he or she ends up getting promoted. This position provides a promotional training opportunity that is awesome for an aspiring officer. Beyond hitting the books, when it comes to the tactical and in-basket personnel assessments, they've been there, they've done it, and they've worn the t-shirt. They usually do well at promotion time. The addition of this position to our department has been wonderful and well worth it. The chief's aide is a position well worth bringing back to the forefront.

This one's easy!

Keep the guys' best interest at heart, always. You can ask anyone in a white shirt in Lewisville, "What's the number one rule on the administrative side of things?" They'll tell you that the guys come first. Period! They come first with our decisions, our budget (budget for what's really needed and for what's best), our programs, and the whole thing. Sometimes the troops have a hard time understanding that and don't believe it—hey, I didn't earlier in my career— but it has to be that way if you want things to go well. It's hard to screw up if you have the best interests of your people at heart.

I used to have a hard time understanding where one of my bosses was coming from and how he could be so suspicious of his guys and so paranoid. He was nasty to them, saying things like "They're just going to do this" or "They're just going to do that" and "If this were the private sector, I'd fire half of them." The first thing I realized was that if he was in the private sector, someone would have clobbered him or he'd be in jail. The second thing I realized was that the terrible things he seemed to expect from his people were the kinds of things he would have done if he were in their position. One of my earlier company officers, Lieutenant Bill Allen, called it "a thief marking his own tools first." The reason that it's so easy for people like him to come up with all of that nonsense is because that's their "M.O." Bill told me not to get frustrated and reminded me that with some people it's like yelling at the rain. They just don't get it. It's easy to avoid being like that. Just remember that the guys come first.

Be careful what you wish for

Be especially careful what you wish for when looking for a new chief. Often the troops will ask management for a chief who knows the operational side of things as well as the administrative side and one who will communicate and be available for the guys. They say they want someone who will show up in the firehouses once in a while and show up at calls. But when they get it they change their minds or have second thoughts. I know. I used to show up at some minor calls or at first reports of structure fires that really weren't serious. The question was raised about trust; was I there to watch every detail because I didn't trust them (fig. 4–10). That

Fig. 4–10. Having the chief respond to incidents helps show support for the command staff and firefighters

couldn't be farther from the truth! My wife would ask me, "Why are you going on that one. It's just a sparking electrical outlet." I would explain that I haven't seen "C" shift in a while. It was a great time to visit with the troops. Our department has our trust and then some. They don't need a babysitter. Going to a minor incident was just a great way to see some of the guys that I hadn't seen in a while. That's all.

There is a department nearby looking for a new chief and it has been interesting to hear some of their guys saying that they aren't looking for a new chief who is young and trying to make a name for himself by coming in and doing a whole lot. They also don't want an old chief who is coming there just to retire. They want, believe it or not, someone kind of like their old chief, even though they disagreed with him from time to time. He was a pretty progressive guy who loved the job. I wish them the best!

The following are some suggestions and advice I've gotten from some of my mentors over the years that I consider very good road maps to follow and which might help you sidestep some land mines.

1) Whatever you do, whatever you say, make sure that the safety of your people comes first. This is for both *in* and *out* of the firehouse. Provide the safest operation that you can.

2) Don't let tradition hinder change, but remember that there are a lot of traditions in the fire service that are great and should be carried on!

3) Be a good leader first. Be their buddy second. They really want someone to lead them. Have the courage to lead. It takes a strong person to make tough decisions and to stand up for what is right. Don't confuse what you think is right for what is *really* right. Do the right thing!

4) Surround yourself with good people. No one can do it alone. Find good people and bring them into your camp. Find people who have the same core values, the same foundation that you have. It's easy to do good things with great people. Just remember that the best thing about having great people is that they're great; the worst thing is that everyone else wants to steal them away from you. That's why your mentoring program and your ability to build tomorrows leaders become so important. Start filling your "people staging area" now. Get them ready to move up to the next alarm. Besides, you owe it to them.

5) Keep it simple. Try not to make things so complicated that no one can figure them out, because you'll end up with a big mess, kind of like dropping a bag of marbles on the floor and watching them go in every direction. Sometimes the best way to get the job done is right in front of your nose and is easy to grab hold of.

6) Lead by example with your day-to-day activities, your uniform, your tactics, and your personal life. Be a good role model for others to follow.

7) Don't credit grab! Give credit where credit is due. Even when it's your idea and it goes well, tell them it came from the troops. You know, it's always been amazing to me that there are a lot of chiefs out there who think that they're the reason that their department is successful. They don't realize that it's not because of them, but because of their firefighters. The firefighters are the ones making it happen. They're the ones out doing the work. So spread the wealth. It looks good on everybody. Bang a drum for the troops, don't bang on the troops. Go out and say good things. BRAG ABOUT THEM! It's okay, really.

8) Focus on what's going well, on how good they are, and on their accomplishments. A good example comes from one of Notre Dame's great football coaches, Lou Holtz. Before some of their toughest games he would get in front of his team in the locker room and ask each player why they were going to win the game. He would write down the reasons given by the players until the board was filled up. All he did was remind them, or let them remind themselves, that they were good and that they could win. Sometimes we need to remind our people how good and special they really are.

9) What kind of a chief would you want to work for? Remember what it was like to work for a bad boss. Stay current, stay in touch, constantly re-evaluate how you're doing, look for ways to improve, and remember what "open door" really means. Are you listening? And your idea isn't always the best one. They've got some great stuff.

10) When you can't figure out why the troops feel a certain way, why they are saying what they are saying, or seem to not understand what you're trying to do, especially with the budget, remember that you were there once and probably yelled the loudest. Always keep in mind that "leaders enforce values, while managers enforce rules." Good chiefs will support their people and let them go. Do this and I promise you that good things will happen.

11) Remember to stay in touch with the troops.

5
OUR TWO FAMILIES

The person who coined the phrase "blood is thicker than water" in an effort to describe the closeness and loyalty that the members of a family share obviously never got to experience the closeness and loyalty of the American fire service family. Granted, your family at home is going to rank higher, and rightfully so, but the fire service family has to run a very close second. Nowhere on earth does there exist another organization that has the closeness and love for one another that the American fire service has. I know, I said love and it sounds mushy, but it's true.

No matter where you go, when you see someone wearing a Maltese cross on his or her chest you feel that closeness instantly. You know what I'm talking about. When you talk to someone you meet while traveling, or at a wedding, or a ball game, as soon as you realize that he or she is a firefighter, you feel at ease. It's as if you've known each other for years (fig. 5–1).

Fig. 5–1. Whatever the event, the "our home is your home" concept goes a long way with both firefighters and citizens.

Now I also understand that we all have that "crazy" uncle that you don't want to let out of the house, but the fact still remains that *we are* a very strong family. In what other organization do you find people lining up to provide help and support as soon as they learn that there is someone in need or having a rough time. It doesn't matter whether it's a sick child, family problems, or any other kind of difficulty: they're there.

There is also a bond and a trust between the public and the fire service that is second to none. Maybe that's why you can't turn the television on without seeing a commercial with a firefighter or fire truck in it. There are Fortune 500 companies that would give anything to have the marketing appeal of the American fire service family, especially when, at times, the American family is struggling with issues like divorce and abuse, as well as so many other challenges. Look at all of the advertising that makes use of the positive image the fire service family has. They use us in every kind of add from headaches to hemorrhoids. I know it might appear that I'm bragging a little but so what? The fire service deserves it.

Defending Our Family

Once in a while, we need to drop some gentle reminders to our guys about defending our family. It's hard to defend your family from those on the outside when someone is attacking it from within. What I'm referring to is when we hurt each other with what we say about each other and what we do. I'm not trying to get warm and fuzzy, but aren't we supposed to take care of each other? We're supposed to be there for each other through thick and thin. Just give them some gentle reminders.

Don't Allow Anyone to Tarnish Our Image

Don't allow anyone to tarnish our image. That means that we should keep each other in line and come down on someone when they cross that line. We've got an awesome family, but we're human. Once in a while someone is going to make a mistake and that's okay. They're called "on accidents" not "on purposes." Occasionally someone is going to screw up and commit one of those "on purposes" when they really should have known better. When that happens, we need to step up and let them know it's not right. If you let someone go too far it hurts all of us. Do you want to try to explain to the public how a firefighter could commit arson (something we read about way too often). That's a difficult task, seeing as how we're supposed to put fires out! It's pretty simple. Our family is not supposed to break the law.

> "They're called on accidents, not on purposes."

Hiring the Right People

So how do you work to prevent your guys from hurting our family? Start by hiring the right people and letting them know just what our expectations are of them. Tell them right away. Tell them that we don't steal, we don't lie, and that we don't put up with anyone who betrays our family. Yes, it goes back to being a good fire service brother or sister. The concept of brotherhood defines a value system that we need to abide by, one in which brothers stand by one another and stand up for each other, for the fire service, and for our family. It does not, however, mean that you take advantage of each other and play on the whole brotherhood thing for personal gains.

Being a brother means I will do everything I can for you, but it also means that you as my brother would never ask me to do something that would risk my own family's financial security. As a brother, I will do what you need, but when it comes to hurting my family, you'll understand and be able to reason with that. Remember, doing the right thing is not the same as doing the wrong thing and trying to make it look right. It means doing the right things for the right reasons. Hurting someone's family doesn't make it right.

The I.A.F.F. and the F.O.O.L.S.

There are two organizations that exist to promote the brotherhood and thank God they do. The IAFF (International Association of Fire Fighters) does this every day for their brothers and sisters. The FOOLS, or the Fraternal Order of the Leatherheads Society, is an organization that works to promote the fire service, provide high-quality training to firefighters, protect firefighters, and promote the brotherhood of the fire service. The FOOLS' motto states, in part, Protect the Brothers (PTB) and Everyone Goes Home (EGH). Well said, brothers!

> "Protect the Brothers (PTB) and Everyone Goes Home (EGH)."

During times when integrity is a struggle in the private sector and family values are being thrown aside by some, we have to step up and protect our own. Maybe we ought to let the fire service family serve as a role model for others to follow. I think in many places it already does.

Our *first* family

Our first family, *the one at home,* is where it all has to start. The only way we can build and strengthen our second family, or extended family, is to work towards building a strong foundation at home. Without a foundation, a strong base, our value system begins to crumble. That in itself is a huge weakness and is a sign of the beginning of the end. For example, take a look at someone in the fire service who is struggling, always in trouble, or one who has chosen to work against the whole brotherhood thing. Look at someone who, at times, has hurt a fire service brother or sister. Nine times out of ten, when you take a closer look, you'll see that the person is lacking a foundation. They lack a values system and they lack it at home. It all starts at home. Help them to develop those values, to learn to live by them, and then you will begin to strengthen that foundation.

Setting Your Priorities

Building upon that foundation also means keeping your priorities in line. I love the fire service more than life itself, but my family comes first. With all of the demands our profession can place on you, such as schooling, training, continuing education, studying for promotions, and even just the time it takes to be good at what you do, it can be easy to forget what's waiting for you, patiently, at home. And when you look at the big picture, isn't family what it's all about? Remember to take care of what's at home. If you're not good at that, you're probably not good at work and may be struggling there as well.

Consider Ride-outs

In Lewisville, we try to encourage the kind of atmosphere that helps our folks with their "first family." Earlier, I mentioned the importance of having awards ceremonies in which the whole family can participate. Hold hiring and promotional ceremonies that bring families together. Encourage family members to visit the firehouse more often and make them feel welcome and at home. We take it a step further, allowing the spouses and children of our firefighters to witness and experience what dad or mom does at work. Not a week goes by in which you don't see one of our member's kids riding out.

Once in a while the question of liability comes up. We have rules that define what a rider can and cannot do while riding out with a firehouse or company and everyone who rides out with us signs a waiver (fig. 5–2). We believe that the benefit to our members of this particular program outweighs the possible risks. Maybe by accepting this type of liability we can help reduce the "liability" some of our troops face at home. We hope it might help keep a few more families together. We have department and company t-shirts in children's sizes for the kids to wear and even have some small sets of turnout gear (fig. 5–3). We really want our firefighters' families to feel that they are part of the fire service. This is just one small step toward getting that done.

Fig. 5–2. Having those from the "first family" ride out and attend department functions helps foster a better understanding of what mom or dad does at the firehouse.

Fig. 5–3. The kid-sized bunker gear has been a big hit with our kids as well as visiting children

Here are just a few ideas to help keep both families strong.

- Remember where your priorities are. Take care of the "first" family. It's a great way to start taking care of the "second" family.

- Remember what the brotherhood is *really* about. It's a 24/7 commitment.

- Just as important as taking care of the things at home, remember to never betray the family at work.

- Establish a 100 Club, or as we did here in Denton County, the Heroes of Denton County. Both are organizations that are set up to assist the families of firefighters and police officers who suffer a loss due to a line of duty death. There are many similar organizations already established across the country. But, unfortunately there are still areas that don't have that kind of support and assistance program in place.

- Implement a good program for handling a line of duty death. Whether you model it after the National Fallen Firefighters Foundation "Taking Care of Our Own" program or a similar program, set it up now, so that if, God forbid, you suffer a line of duty death, you're there to take care of the family. When it comes to providing the respect and dignity that a fallen brother or sister and his or her family deserves, there are some really good resources out there to model after, such as the program Chief William Peters put together for *Fire Engineering*. Another great program is furnished by the International Association of Fire Fighters with their *Fire Fighter Line-of-Duty Death and Injury Investigations Manual* or another one that is in place with the Denton (Texas) County Fire Chiefs' Association and their Line-of-Duty Death Investigation and Firefighter Funeral Procedures Policy.

- Remember to be fair, objective, and honest at work. Stay open, encourage diversity, and just be nice.

- Remember what the 9/11 and memorial stickers on your helmet are for. NEVER FORGETTING means NEVER FORGETTING (fig. 5–4)!

- And probably the most important of all: make good decisions and think before you act. Make the right decision because it is the right thing to do. Decisions based on emotions are almost always wrong and can hurt or get people in trouble. Decisions that are made on good solid information and common sense usually go a long way and give better results in the end. Remember that each decision you make is going to impact someone in one or both of your families. We teach our kids not to drink and drive, to avoid drugs, and not to get into a car with someone who is under the influence. Once in a while, we need to remind the people in our second family as well.

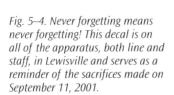

Fig. 5–4. Never forgetting means never forgetting! This decal is on all of the apparatus, both line and staff, in Lewisville and serves as a reminder of the sacrifices made on September 11, 2001.

Managing the Consequences of Other People's Actions

As you pass from day to day, life will throw you a couple of turns now and then. The choices you make can lead to success or set you up for failure. Make good decisions and make good choices. Remember, we're in the business of managing the consequences of other people's actions. Sometimes we need to remind ourselves of that very same thing when it comes to actions we take. Remember what's at stake. Making good decisions, taking care of *both* families is the only way we can keep the foundation that supports our values, from crumbling.

6
SWEATING THE SMALL STUFF

There are people out there who will tell you that it's not important to be concerned about all of the little things that life presents you with. They may say that it's not a big deal and that you shouldn't "sweat the small stuff." They will explain that worrying about all of the small stuff, things they sometimes refer to as *insignificant*, does nothing more than complicate your life and create a whole lot of stress. But I think that it's often the small things that we overlook, or fail to pay attention to, that cause us the big problems in life. I'm not recommending that you become a worrywart or get nuts over everything, but I do recommend taking better care of the little things.

When you think about it, life is really not that complicated. It's rather simple, if you let it be, and as long as you don't allow technology to drag you completely to the other side. There's a lot of good stuff out there, but keep in mind that you have to breathe, blink your eyes, and relieve yourself. Your Palm Pilot can't do that for you—at least, not yet. When you start to think about it, with all that has happened to us, maybe its time we started worrying about the little things, the not-so-technical things, again.

There are a lot of books out there that exist to tell you how the author thinks you should get organized, live your life, and do your job. I'm not talking about the books you should be reading, like John Salka's leadership book, *First In, Last Out: Leadership Lessons from the New York Fire Department*, or John Norman's *Fire Officer's Handbook of Tactics*, or any of Skip Coleman's books, I'm talking about the books that tell you not to sweat the small stuff, because it's all small stuff (fig. 6–1). I just can't agree with that when it comes to our job in the fire service. In our job, we have to sweat the small stuff. It's what keeps us alive.

Read reports. Read articles. Look at the facts that led up to a firefighter fatality or injury. The vast majority of the time you'll see that it was the small stuff that got us into trouble in the first place. "Saving Our Own," "Get Out Alive," and all of the firefighter rescue programs are great, but maybe if we trained really hard on the little things, the small stuff, the basics, we wouldn't get into

Fig. 6–1. Chief John Norman

the kinds of jams that mean we need to use what is taught in those classes. It's often the small stuff that kills us, so before we can be good at all of the fancy stuff, we have to be good at the basics. I know we keep hearing that, but it's true. We have to know our job! There are no excuses. We have to be good at what we do. If we're not, we're just a short step away from disaster. Then what will we say?

The Line-of-Duty-Death Book Report Drill

We've used a drill for some time now that has had some great results. We assign a firehouse shift to a past firefighter fatality. We give them as much information as we can regarding the incident, and then ask them to do some basic research. We ask them to identify what they feel were the contributing factors that led to the death and to apply those factors to the way we operate and what we do here in Lewisville. Are there any similarities as to how we operate? Are we traveling down the same path on our way to disaster? Then they are asked to present what they have found to the rest of the shift, either each firehouse separately or together as a drill. The presentation doesn't have to be fancy and doesn't have to take hours to present. What we end up seeing are some pretty good, well-thought-out presentations coming from all the different ranks, not just the company officer. They discuss lessons learned and make comparisons to how we operate. What's interesting, though, is that after they do a few of these reviews, they start to realize that most of the contributing factors carry over from one fatality to another. Most of it is the same stuff, killing firefighters over and over again. It's a great way to demonstrate the importance of sharing information and handing it down year after year. Only by sharing information will we stop killing firefighters in the same ways over and over again, and hearing that they didn't know that it could happen to them, when in reality it has, several times, just in other places.

When you look at the leading causes of firefighter fatalities and injuries—lack of command and control, failure to follow SOPs, lack of an accountability system, poor communications, and some of the others—most can be broken down to small things, like failing to follow SOPs. Having a set of standard operating procedures allows all of the players to play from the same playbook. Telling everyone to "go long" just doesn't work for us. We need something specific to operate from. In a world now full of "control-alt-delete" and all of the "hyperspace" stuff, we're forgetting about some very important things.

The Little Things

One little thing that comes to mind immediately is basic search techniques. Thermal imagers are great, but what do you do when you don't have one, or when the battery goes dead, or when you drop it in the smoke? You still need to know how to perform a conventional search. You still need landmarks such as an outside wall or a sofa, and you'll still need to get out eventually. Do you know how to stay oriented enough and could you find your way out if you got separated from your partner? My best friend, FDNY Battalion Chief John Salka, teaches a program on staying oriented in a burning building and how to get out if you get lost (fig. 6–2). Toledo Deputy Chief Skip Coleman teaches a class on the "oriented room search," a program on how to stay oriented while performing search and rescue operations (fig. 6–3). Both are popular classes because losing orientation is a way a lot of firefighters get into trouble.

Fig. 6–2. Chief John Salka

Fig. 6–3. Chief Skip Coleman

And how are you at advancing hose? How many times have we heard our good friend Andy Fredericks say that getting that first line into place as quickly as possible is what's going to make or break your incident (Fig. 6–4). He says that being proficient at hose advancement will make for a successful attack. Don't wait until you're inside to find out that you don't have water, that your line is kinked on the stairs or out on the lawn or wedged under a door, or that you didn't clear the hose bed completely when bleeding off your line before going in would have prevented that.

How much do you know about little things like building construction and fire behavior? If you want to be good, you need to know fire behavior and building construction. You have to know how the building is going to react with the fire and how the fire is going to react with the building. Have I mentioned that part yet?

Not too long ago, while a handful of our firefighters were cleaning their rig, I asked them who Frank Brannigan was. (I was trying to make a point to a couple of my chiefs.) Well, the firefighters didn't know who he was. I gave them a hint; I told them that he had written a book. One answered, "You mean the one on safety?" I said, "No, that was Vinny Dunn's book. Another great one but that's not the one I'm talking about." I gave them another hint.

"He writes a bi-monthly column in *Fire Engineering* called the "Ol' Professor." They looked at me. I said, "They are smaller articles on building construction. You know, information about some of the small stuff that saves our lives." By the way, next month's drill: building construction.

Have you read Tom Brennan's monthly article called "Random Thoughts" (fig. 6–5)? It's the one all the way at the back of *Fire Engineering* magazine. It's usually about a page to a page-and-a-half long, nothing fancy, and covers small stuff that makes a difference. I have a funny story about Tom Brennan. A few years ago, I had the privilege and honor of giving a speech called "Pride and Ownership" at both FDIC in Indianapolis and FDIC West in Sacramento, California. In Indianapolis it went well, but in Sacramento, I ran into a snag. About five minutes into my speech, I looked out into the audience and saw Tom Brennan sitting in the middle of the second row. Well, I vapor-locked. I mean, it was *Tom Brennan*. My sector C slammed shut. I could feel my pulse pounding near my temple and my eye started to twitch. But then I remembered that he was one of the reasons that I was up there talking about pride and ownership in the first place. Like some of my other mentors, that's what he's all about. But the neat thing about Tom is that he has

Fig. 6–4. Lieutenant Andy Fredericks

Fig. 6–5. Chief Tom Brennan

the ability to take things that a lot of people have made very difficult and complicated and break them down so that everyone can understand them. He shows you the little bits and pieces that all add up to the big picture—you know, the small stuff. The rest of the speech went fine.

The Three "F's"

Another "ol' professor" of mine, Chief Eddy Enright, taught me about the "3 F's." (Well, it's actually four F's if you spell firefighter as two words, but I like to spell it as Bill Manning did, as one word.) The three F's are Firefighters, Fire Apparatus, and Firehouses. It seems kind of simple, just those three things, but there are some really great things to think about with each one.

"Firefighters, Fire Apparatus, and Firehouses."

The First "F"— Our Firefighters

Look at our firefighters, the first "F." Are we taking care of each other as we should? Why did it take the events of 9/11 to get us all calling each other brothers and sisters again? It's been nice not hearing that whole East coast versus West coast and North versus South rivalry, hasn't it? We've come together as a team, *one team*. And the hugs feel better than ever.

But are we taking care of the small things, like our uniforms? Do you know what your badge, insignias, and uniform all stand for and where it came from? What about your turnout gear? Is it ready and in good shape before you go in? Do you have an inspection program? Before you go in, do you take a look at your partner? Is he or she as ready as you are? Does everyone have their hood on (fig. 6–6)? Is your team ready? Did you wear your gear to the scene? Let me clarify, did you put your gear on before you left the firehouse? I remember an extra-alarm incident we had at our hospital during which we had a mutual aid truck company show up. We needed to put them to work pretty quickly but had to wait while they put their gear on and got ready. It was frustrating because they left their firehouse knowing that they were going to a working fire. We're the first ones to complain about the lack of protective clothing, or when what we have isn't good enough. We have to put it on when we're supposed to.

Do you bleed off your line? Do you know the lengths and amount of hose in your hose beds and how to get it there? Do you understand and know your nozzles and what they will and will not do for you? Can you work with a bigger line when you need to and are you efficient when deploying it? The 2½-in. attack line is not difficult to deploy and use if you train with it. Do you use door chocks to block open a door while you're searching a room and to keep a door from closing on you and your hose? Is your flashlight working? Do you even have a flashlight? As you walk around the exhibit floors at the next show you attend, after you buy all of the t-shirts you want, buy yourself a good flashlight. Carry two: yours and the department's. Don't rely on the department's flashlight alone. Don't be stubborn. Your life is worth the investment. As Chief Freeman has always said, "Don't let them put on your tombstone, bastards wouldn't buy me a flashlight."

Fig. 6–6. The Captain in this picture is "dressed for the party" and ready to go.

Is your portable radio working and ready to go, with a charged battery? Are you working to ensure that every firefighter has his or her own radio? If you have one, do you know how it works? What about your PASS device? Is it working and do you know how to operate it? We used to call them slacker devices because if they went off it meant that you weren't doing any work. But, to put it bluntly, do you realize just how many firefighters would be alive today if they had just turned on their PASS device and made sure it worked?

RIT Doesn't Suck!

Do you believe in rapid intervention crews and do you have one ready when you're operating on your scene? Not everybody does yet. A deputy chief from a large metropolitan department is out there lecturing and one of his topics is that rapid intervention doesn't work. He says, "Rapid intervention sucks!" Maybe it does the way he's doing it, but there is case after case after case of documented saves made by rapid intervention teams that have gone in and rescued a fellow firefighter. I wonder if the firefighters who were rescued think rapid intervention sucks. I wonder if their families do!

Do you carry side cutters or a Leatherman tool to help cut a firefighter (or yourself) out of an entanglement with wires and can you get to them with your turnout gear on? What about your accountability tags? Do you use them or do they just take up space on your helmet or turnout coat? Accountability isn't just the responsibility of our officers and it's more than a board with Velcro on it; it's the responsibility of each and every one of us. Don't we owe it to each other and to our families? Whatever accountability system you have, are you using it? Keep in mind that the system that you are using is to accountability what a Halligan bar is to forcible entry. If you don't use it or don't know how, it won't work.

One more little thing: how about your self-contained breathing apparatus (SCBA)? Do you know it inside and out? It's probably one of the most important things you're going to bring to the fire. Another thought to ponder: Approximately 85% of SCBA malfunctions are a result of operator error. Get to know your SCBA on a first name basis. What programs do you have in place to force you to become proficient with your SCBA? You owe it to yourself and to your partner to be great at it.

Are your tools ready and do you take care of them? Do you know how to use them? Showing up on the fireground without your tools is not a good thing unless you're good at magic tricks. And get rid of the closet pike poles. You can't force your way in with them and sure can't force your way out with them. To be honest, they're just not a good pike pole. I've seen firefighters standing on top of chairs trying to reach the ceiling with closet pike poles. I'm not really comfortable reaching up over my head through the smoke and hoping to pull something down and possibly onto my head. Grab a tool that will do some work for you and that can get you out of a pinch if you need it to.

Your People Are a Reflection of Your Own Self Image

How's the attitude in the firehouse? Are the guys interested in the job or are they just hanging around waiting for the eagle to relieve itself? Remember, captain or lieutenant, the attitude in your firehouse is in your control, as is just how well the guys are going to do. Do you have a good hydrant inspection program? Do you know where your fire hydrants are and do you know if they work or not? How about your ground ladders skills? Are you good at throwing ground ladders, not only on the side of the firehouse, but also out where we do our work, with the obstacles and challenges that we see when we have a job?

> *"You follow ugly kids home, you're going to find ugly parents."*

We're Losing on the Streets, Big Time

Are we addressing traffic control at accident scenes and are we doing everything we can to get there (and back to the firehouse) safely (fig. 6–7)? When you are in your rig, whether you are responding to a call, returning from one, or just out and about, are your personnel wearing their seat belts (fig. 6–8)?

Fig. 6–7. Scene safety and apparatus placement are critical to firefighter safety at motor vehicle accidents.

Fig. 6–8. Reflective vests, good reflective trim on turnouts, and a watchful eye are paramount for MVAs and other incidents that place you in or close to moving traffic.

Seat belts save lives

How many times have you been at a serious motor vehicle accident and found out that the person in bad shape or dead, the person ejected from their car, wasn't wearing a seatbelt. Have you heard the guys saying that it could have been different if the victim had been wearing a seat belt and then watched the firefighters get back on their rig and pull away without putting their seatbelts on? Or—and I just can't understand this one—they get back on the rig and ride on the side- or tailboard. We've been so far down the road on this one that it's amazing to see that we still have firefighters riding the rear step. What's it going to take (fig. 6–9)? Are you using a back-up person? Is there someone behind you and off to one side, visible in rear view mirror, backing you up so that you don't hit something, or, worse yet, somebody. What's the excuse for running over one of our own?

Fig. 6–9. A great seatbelt program for the fire service, "Everybody Goes Home" is already having tremendous results in seatbelt awareness and safety.

Once again, how's the leadership and attitudes of the firefighters? Do they know that their officers care about them? *Do* their officers care about them? They had better. Our company officers are where it all happens. How well are you taking care of your guys? Are you mentoring and building tomorrow's leaders? Are you preparing them to move up? Remember to have a succession plan and to build tomorrow's leaders. Remember how your attitude and leadership can affect the safety and well-being of your firefighters.

People Staging

It's important that you're taking care of what we referred to earlier as "people staging." Remember that you're building your own legacy. Are you walking the talk with your values and lifestyle? Are you helping them with their promotional studies and study habits? How well are you preparing your men and yourself for the future? Is their attitude in check? Chiefs and officers, are you letting them be firefighters? What about that other little thing called communication, from shift to shift and at roll call? When you don't have roll call, you miss out on one of the most important times of the day. I'm not talking about the rumors that bounce back and forth but about the information that is exchanged concerning what's going on and what happened the day before. I used to find it amazing that in some firehouses, when you went for a visit or a cup of coffee and asked them what they thought about the fire that the shift had handled the day before, they didn't know anything about it. How can you not know about a job that happened in your own district the day before? Roll call is also a great time to size up your people.

> "I always found it funny that the shift before you are slobs, and the shift after you are the nitpickers."

Now, about the rumors spread at roll call. I've said many times before that this is the best job in the world. Can you imagine what it would be like if you didn't have to worry about rumors or worry about what the other shift was saying? I always found it funny that the shift before you were the slobs and the shift after you were the nitpickers. That's only until you change shifts and realize that your old shift were the slobs and the other shift were the nitpickers.

Seriously, can you imagine how much better this job would be if we were nice to each other all of the time? How many times have you heard "Why do we have to wait for a fire to all get along?" Be nice, say nice things, and try to get along. Be a brother. The bottom line, again, is that we need to take care of each other. Officers, you need to take care of your guys. There are too many people on the outside who don't care about us. Those are just a few more of those small things.

The Second "F"— Our Fire Apparatus

The second "F" is for fire apparatus. Are we taking care of our rigs? Are they ready to go? Whether you work shift or come back for a call as a volunteer, will your rig start when you need it? If your shift starts at 7:00 AM and you catch a call at 8:00 and your rig doesn't start, what will be your excuse? Being "dead in the house" an hour into your shift doesn't cut it.

Do you know what's on your rig, or are you one of those guys who run around the rig slamming compartment doors when you're looking for a tool. Knowing what's on your rig is not going to happen by osmosis.

You're not going to absorb that information through the compartment door. Do you know how to use all of the equipment on your rig? Do you know what job is assigned to the particular seat you're riding in? Do you know what your task will be when you get to the scene? Lastly, does the rig look good? These are small details, but they're important.

The Third "F"— Our Firehouses

The third "F" is for the firehouses. Guys, it's our home (fig. 6–10). Do you treat it as if it were your own? It's supposed to look good. In the past, when going to teach at a fire department, I used to try to arrive a day early so that I could become familiar with the department, meet the guys, and try to get a good read on the department. A friend of mine, Chief Bennie Crane (fig. 6–11), shared something with me back then that I've never forgotten. He said, "All you have to do, Rick, is get there about ten minutes earlier for class then you had planned. Take a look at the rigs. See how their hose beds are finished. Take a look at their tools, how they have their gear

Fig. 6–10. If it looks like a firehouse, smells like a firehouse, and tastes like a firehouse, it is probably like this Lake Cities, Texas, firehouse.

laid out, and look around the firehouse. Then take a look at the guys. Are they into their jobs, do they have that love for the job, or are they too busy reading the NASDAQ report or watching the fishing channel? Is their number one job still the number one job? Doing that will tell you all that you need to know about the department and the guys. There's no real science to it, they'll tell you everything about themselves without you having to ask one question."

In Lewisville, one way we try taking care of the small stuff is in our response to incidents. We try to live by four very simple rules.

Fig. 6–11. Chief Bennie Crane

Rule number 1

Every time we go out the door we're going to a fire. It doesn't matter if it's "just that fire alarm again." When we leave, we're dressed, we're ready, we're in the right frame of mind, and we have our game faces on. When we do that we reduce the odds of getting snookered and hurt. We fight off complacency.

Rule number 2

There's no fire unless we say there's no fire. We'll take into consideration what the police officer or a civilian is telling us, but we'll make that final decision. We don't like surprises.

Rule number 3

There's no one in the building only if we say there's no one in the building. Once again we'll listen to the information being offered *but we still search*. We don't want to miss the neighbor that came in the back door to help when everyone else was going out the front, or the kid who snuck back home and didn't tell his parents. We don't rely on people whose judgment may be impaired because they woke up in smoke. Some of these people might not remember who was in the house and may believe that everyone is out.

Rule number 4

The fire's not out unless we say it's out. It's not above us and it's not below us. When we ensure that it's out, we don't see fires rekindle and cause us embarrassment.

Follow those four simple rules and you're less likely to get yourself or your people hurt. While abiding by the above four rules we also keep in mind our risk management statement, which I'm sure you've seen before (fig. 6–12):

1) We will risk our lives a lot, in a highly-calculated and controlled manner, to protect a human life.

2) We will risk our lives a little, in a highly-calculated and controlled manner, to protect property.

3) We will not risk our lives at all to protect lives or property that have already been lost.

Maybe you don't always have to sweat the small stuff, but taking care of the little things can often keep them from becoming big things.

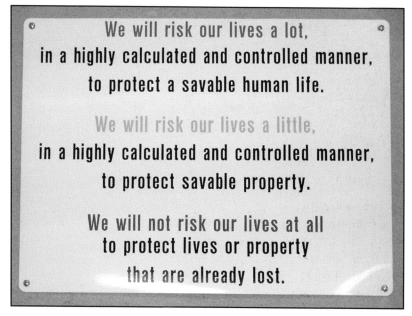

Fig. 6–12. This risk management sign is posted on the way out of and on the way into all of the apparatus bay floors in Lewisville.

7
CHANGING SHIRTS: THE PROMOTION

One of the many opportunities available in our great fire service is the ability to move up the ladder if one chooses to do so. The opportunities are there whether you work in this wonderful profession as a volunteer or as a paid firefighter. Preparing for promotion, and if you're fortunate, getting promoted, requires some planning, or "promotional pre-planning," if you will.

Promotional opportunities can have strange ways of presenting themselves. You could be lucky enough to work in an area that is rapidly growing and is building firehouses and hiring firefighters to keep up with that growth. Along with that kind of growth come promotions, and if someone isn't staying awake (and it has happened), a fire department ends up playing catch-up, trying to fund firehouses, apparatus, and personnel that should have been planned for in the first place. A friend of mine is going through this right now. He replaced a good chief, but his predecessor didn't do much planning for the future. My friend has to acquire more firefighters, new firehouses, and new apparatus, and is in need of it all at once. On the other hand, you could be sitting in a firehouse that hasn't seen a promotion in years for a variety of reasons. If you're in the latter, it can be hard to stay fresh and in the promotional preparation mode.

Don't Become Stagnant

Time and time again, we see firefighters and officers decide not to prepare and enter the promotional process because they don't see any promotions happening within the list's dates, only to see them get caught by surprise when someone retires unexpectedly, new positions are suddenly approved, or someone wins the lottery. They're always left wishing they had prepared and taken the exam (fig. 7–1). The best way to stay sharp and fresh is to prepare, enter the promotional process, and see what happens. About the worst thing that could happen is that you could get better and better at taking the exams while staying current and on top of things. Don't fall into a rut and remain stagnant.

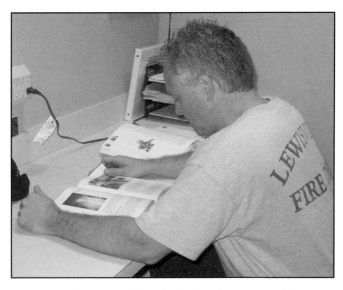

Fig. 7–1. Studying, whether there is an open position or not, will keep you current and sharp.

The Decision to Promote

Upon deciding to enter a promotional process, you'll need to ask yourself several questions.

First, why do I want to promote?

If you're a paid firefighter, is it for the money? Is it because it seems like the thing to do? Or is it to make a difference? I understand that if you're paid the money might be an important factor, but hopefully the last one will be the biggest reason. The desire to make a difference and to contribute and the knowledge that you have something to offer should be the driving forces. When people who don't feel this way get promoted, they become officers who don't accomplish much and seems like they're just hanging out.

In the decision process, don't get all wrapped up in the "I'm only taking it for the experience" stage. You are either serious or you are not. If you are serious, give it your all and do the best you can. If you're not serious, you may experience the process but you won't learn where your strengths and weaknesses are.

Can I do the job?

Hopefully your answer will be yes, and a loud yes. If you're not sure, maybe it isn't your time yet. You don't want to get promoted into a position that you weren't ready for. Talk to one of your mentors and see what he or she feels about your situation and whether you should go for it or not.

We've all seen the firefighter who has been on the job for about a year and has seen a little action (and I mean a little) who looks up front at the driver's seat, usually at someone that has been driving for a while, and says, "I can do that. That's not so hard." And when that guy eventually gets into the driver's seat, it's not long before he's looking at the officer sitting to his right and saying, "Hey, I can blow the siren and talk on the radio. That's not so hard." Once he makes it into the officer's seat he starts looking at the battalion chief and saying (I hope you can see where I'm going with this…), "Hey, that doesn't look so hard. I can do that." Eventually, he's looking at the chief of department, saying, "Hey, that doesn't look so hard, either." It's not bad to think you can do it better than the other guy. Just don't forget that it takes preparation and hard work to get there as well.

You also don't want to be the guy who gets on the "train of life" saying that all that matters is getting to the next station. (He's the kind of guy who says, "If I get this firefighting job, that will be it. I'll be happy for life.") But as the train gets close to that station he starts looking ahead to the next station and thinking, "Well that's not too far. I'll just hurry past this one and get to the next one. Then that will be it and I'll be set." Then as each new station approaches, he gets the same feeling and just keeps blowing by, not experiencing much, and, worse yet, forgetting to stop and enjoy life for a while. People like him end up asking themselves at retirement, "Where the heck did my career go? Boy, it sure flew by."

Can I make the lifestyle change required?

Changing your lifestyle to meet the needs of your new job may be easy for some, but for others it can be extremely difficult. Most of us have said something and may even have done some things that seemed like the right thing at the time, only to realize later that in order to take on more responsibility, you have to *be* responsible. You should know, early in your career, that the decisions you make and the actions you take as you rise through the ranks can and will affect those who work with and for you. The things you do early in your career can hurt you for a long time. Early on, make the changes that are necessary to handle your responsibilities. If you were screwing up or were a problem child before being promoted, "changing shirts" doesn't make all of that go away. We've seen plenty of troublemakers get promoted and then immediately come down hard on the guys for doing the same things that they were doing just months ago. Take care of business early. Start early in your career and make good choices.

Do I have the courage to lead?

Being a leader takes courage. Leaders need the courage to make hard and tough decisions, and I don't mean just on the fireground. Sometimes as an officer, a leader, you have to make decisions that are not popular with the troops and many of those troops will probably be your friends. But you have to make decisions based on what's right and not what's popular. In the long run, you'll be happy you did what you did for the right reasons. Making the leap from buddy to boss is not as easy as it seems. Ask anyone who will be honest with you whether anything changed when they got promoted and they'll say "Yes!"

I can tell you first hand that no matter what you do, no matter what your decision is, some of the guys aren't going to be happy. Even when you think you're doing a good job, someone will make you feel that you just can't win. The bottom line is that no matter how hard you try you're not going to make everybody happy all the time. I asked a friend of mine how he was adjusting to being a new lieutenant and he said, "You know, there's good days and bad days." Then he pointed to his collar pins and said "Some days they're trumpets and some days they're funnels." Just remember, there are a whole lot more good guys than troublemakers. We often focus on those giving us grief or those who don't care and forget to take care of the right people. The good folks are going to feel cheated after a while when they think you don't spend enough time on them and that you spend more time on those who are not willing or capable of changing. In the end, if you make decisions based on good solid information and integrity, you'll come out on top.

> *"You have to make decisions based on what's right, and not what's popular."*

> *"Some days they're trumpets, and some days they're funnels."*

Study Habits

Let's talk about studying. Start studying early, way before the promotional announcement. Why wait to start studying until they stick the notice on the bulletin board? If you start early, the days leading up the examination process will be a review rather than the first time you're cracking the binders on the books. Again, it's all going to keep you sharp (fig. 7–2).

Fig. 7–2. Helping another firefighter study for an exam or test is a great way to bring everyone up.

Study with a friend or a partner. A lot of people are afraid that if they study with someone, and in doing so help them, that person will beat them on the exam and score higher. I've always looked at it this way: I need all of the help I can get and if my study partner scores higher than me, so be it. He or she was probably going to score higher anyway. Set aside some time each day to study even if it's just a short time. Try to find a quiet area. It's hard to concentrate at the kitchen table or in the day room. Find ways to test yourself. Truly become a student, especially if you've been out of school for a while. Try to build good study habits and read something about this job every day. It doesn't matter if it's in a book or magazine or on the internet, it's going to help you develop and become a better firefighter.

"Failure to hit the bull's eye is never the fault of the target. To improve your aim, improve yourself."

Provide a Good Process

If you're a chief and have a say in the process, provide a promotional process that is fair and objective. I've had the help of a lot of people, including some of my mentors, and we've done really well with our promotional process. Without getting into too much detail here, possibly the biggest impact has been that we conduct the process on the candidate's day off, after a day off. This allows them to be fresh, well rested and in a good, clear frame of mind. It really doesn't seem fair to put someone through an all-day process when they have just come off shift and were running calls all night while the other candidates were home sleeping. It's a lot more work this way, but it really helps to even things out.

Going to Days

Getting promoted and going to days (Monday–Friday, 8:00 AM to 5:00 PM) can be one of the most difficult changes to make. Not only are you leaving a "shift work" schedule, you're also changing things at home. Now you're home every night. Let me say that again, now you're home every night! That can be a good thing, and we'll discuss that in a little bit, but it is a major lifestyle adjustment at home. I know my wife at times misses the day of freedom she used to get every third day. You know what I'm talking about. My wife and I are best friends, but when you're used to that break every third day, you miss it when it's gone.

Now for some of the good stuff about working days. Yes, you're working Monday through Friday, but you are home every night. You are home for all of the holidays. You miss fewer activities with the kids and holidays with the family. Some people love it. They take to it like a duck to water. However, some do not and the change to days becomes something really horrible and difficult. Under some circumstances the change in shift can strain relationships in the first family, the one at home. For some it may be the reduction in income that occurs because they're not working their side job on their days off anymore. Whatever your circumstances, if the promotion you are going for means that you'll be going to days, make sure your lifestyle can make the adjustment. Doing that will save you from a lot of headaches and hardship.

Here are just a few points to ponder when considering whether or not to "change shirts."

- Form good study habits early in the process and early in your career.

- Prepare to promote early in your career by making good decisions and leading by example now. Don't get to a point where you're playing catch-up.

- If you're worried about the skeletons in your closet, stay away from the graveyard. (Think about it.)

- Set your goals early in your career and then start preparing to reach them.

- Pick good mentors and surround yourself with good people.

- Do a core values check-up. Are yours in order?

- Stay fresh by studying every day whether you see any chances for promotion or not. Don't fall into the regret trap.

- Test no matter what. Enter the promotional process even if there are no openings but don't do it just for the experience. Do the best you can.

- Remember that it takes courage to lead.

- Remember, leaders enforce values, managers enforce rules. It's important to be a manager, but in the fire service the scales are tipped just a little bit more to the leadership side.

> *"If you're worried about the skeletons in your closet, stay away from the graveyard."*

If you get promoted, always remember (and these are just a few):

- Take care of your people.

- Make good decisions based on good information, not on emotions.

- Remember, they're called "on accidents," not "on purposes." We all make mistakes.

- Spend time doing good evaluations.

- Don't stop studying and learning about the job.

- Don't give all of the orders you've been saving for years at the first fire you go to.

- And, probably the biggest: DON'T FORGET WHERE YOU CAME FROM!

8
WHAT 9/11 DID TO US *AND* FOR US

Words cannot describe the loss we all felt and suffered on September 11, 2001. It was an absolutely horrible chain of events that left us in a state of shock and sadness. With hearts hurting, we tried to figure out what had happened, how it had happened, and why it had happened. We worked hard to hold everything together and to hold on to our family, our fire service family. Many in the fire service struggled with feelings of sadness and anger. Just when you started to get a handle on the sadness, you'd think about friends who were no longer with us and get ticked off all over again.

The Public Outcry and Support

The public has always held the fire service in very high regard. When kids are asked about heroes or role models they always mention firefighters. The public trusts us when we're handling an emergency and when we're not. People hire firefighters for odd jobs and for many other services, and when asked by a friend, how was so-and-so at building your addition, they say, "Great. You know, he's a firefighter." So our loss was felt by the public as well and the support they gave us was phenomenal. They consoled us, fed us, and donated money to assist the families of those we had lost. Churches, schools, civic organizations, and whole communities, were there for us. All of us!

A Wake-up Call

Never, in our wildest nightmares, did we ever imagine an event that resulted in such a great loss. After all, that kind of attack, like the one back at Pearl Harbor, just didn't happen in America. The truth is, a few people saw it coming, many never believed it could happen here, and now everyone knows that it has happened and could someday happen again. It was a wake-up call for all of us. Retired General Norman Schwarzkopf said it best when he said, "For decades we've had those two big oceans to protect us, but today technology has taken that away from us. We're as vulnerable as ever." He also added that we got so comfortable that some of our past administrations cut funding to the Armed Forces and hurt the very programs that help us defend our country. Boy, does that sound familiar to us in the fire service.

Where Were the Experts Before 9/11?

Before 9/11, many of us tried to push for programs and classes for our personnel that would inform them about anthrax, small pox, and weapons of mass destruction. Most that were trying to do this were laughed at or told there wasn't any need. Now, finally, we have the classes and training, for the most part, that will allow us to handle these types of emergencies. What is kind of amazing is after the attacks on 9/11, you couldn't swing a dead cat around the room without hitting an expert in terrorism or weapons of mass destruction. My question is, where were they before 9/11? Now don't get me wrong, there are people who are very talented, knowledgeable, and experienced in this area and they are doing a great job getting us to where we need to be, but a whole lot of others came out of the shadows when they heard the money train coming.

The funding is here! Well, some of it…

There's not a fire department anywhere in this country that has a limitless budget, or, if there is, I haven't met them yet. I've never seen a department overflowing with funding and just looking for a way to spend it. The hard, cold reality is that most of us are trying to rub two nickels together to make a quarter (fig. 8–1). We're fighting off budget cuts while trying to get new funding for areas that need it. We need money for little, insignificant things like protective clothing, SCBAs, radios, tools, training, and even, for the paid departments, that unbelievable, unreasonable request of decent salaries for our people so that they can put food on the table (fig. 8–2).

Fig. 8–1. This special response unit will enable Lewisville to respond to manmade events and disasters and handle them more safely and efficiently.

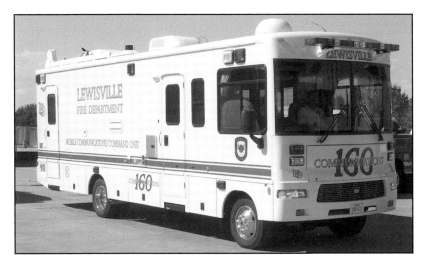

Fig. 8–2. Like the special response unit, this mobile communications/command unit was purchased though a Homeland Security grant and is used by several communities.

For years, a variety of folks have tried to get us the funding we need and to protect us. For a long time, volunteer firefighting organizations, the International Association of Fire Fighters (IAFF), the International Association of Fire Chiefs (IAFC), and editors like Bill Manning have fought to protect us. After 9/11, they had the ammunition to go after some big stuff and they did. But as hard as they worked and as hard as so many others tried, we still didn't get everything that we need. Many fire departments did and still fund programs that tax their budgets or push other areas to be cut so they can fund the "new stuff."

A lot of new grants and funds are there if you know where to look, but you still have to play (or fight) the political game to determine who needs it, who needs it more, and why it's needed. The bad news is that there's still not enough money. The good news is that there is more available than there was before. I often remind myself, when I question the length of time it takes to get a grant approved and then be able to spend the money, that instead of waiting months to get it, we've actually been waiting decades. Think about it. I often ask myself, "Think back before 9/11; how long did it take back then?" Enough said! Waiting a few months more doesn't bother me.

Good politicians—bad politicians

When we talk about the bad and the good that can come out of a crisis, you can't help but get to this one. In an earlier chapter we discussed the fact that, before 9/11, many politicians fought us over funding and on our requests to add more personnel. They fought us hard. After 9/11, BAM! They were gone. Well, sort of. A lot of them lined up to take photos with us, put their arms around us, and promised they would do whatever they could to get us the funds we need to be better prepared.

Now, many did come through for us and we are doing much better. The rest, though, slipped back to where they had been before 9/11. But they'll be back. We'll see them again when there's an election or if we get hit again and they want to line up next to us for another photo opportunity. Then, as time slips by, they'll demand to know why we weren't prepared. They'll start pointing fingers (always away from themselves). We'll remind them about the funding they promised. They'll say it's coming (the funding) and we'll be right back in it again, going around in circles.

Just like the stop sign

Just as it seems we never put a stop sign up at that dangerous intersection until a child gets hit, the fire service never seems to get funding or make changes until one of our own gets hit. Whether it's on the fireground or off, we are too often reactive rather than proactive when it comes to our own needs. This time we have a bigger platform from which to yell and we shouldn't stop screaming for the things we need to protect our people, and ultimately, the public, until we get them. Every time I think about the friends I lost on 9/11 as well as before and after that day, I look for ways to be sure we did not let them go in vain. We talk about how those who have left us will always be with us, teaching us and helping us to go home each day. But they can only do that if we let them! To my mind, their deaths will never have been in vain because they gave us the ability to better protect our people, and, to be honest, a list of good things so long that I can't describe it all in this book.

Just take some time to look at what occurred when hurricane Katrina hit the Gulf Coast. I know that a lot of things went wrong, but a lot of things went right as well. Sometimes the negatives can overshadow the positives. This is the time to stand up and acknowledge the impact that the loss of our fire service brothers and all the others who died on 9/11 had on our ability to handle the problems caused by Hurricane Katrina, and by Hurricane Rita, which hit weeks later. Could the mayor of New Orleans or of any other city stand up on television and order the evacuation of an entire city without the lessons learned from 9/11? Although there were things that didn't happen or go well, a lot of things went extremely well because of lessons we learned on 9/11 and because communities are now planning for handling such disasters.

We're Never Going to See a Disaster Like That Again

There were plenty of people who said, "We'll never have a disaster like we did on 9/11, so why are we doing all of this planning and having all of these drills?" In my own city, several people had to have an attitude-adjustment meeting with the city manager. Thank God he was a forward-thinking boss. But a lot of things went well because of the disaster plans that were created and developed after 9/11. Keep in mind, before

September 11, 2001, there were few cities and fire departments that had a disaster plan, and those that did, were a little weak. We did handle things better and we'll get better learning from the whole Katrina event.

They Made a Difference, Again!

September 11, 2001, was one of our worst days as a fire service and as a country. Our brothers and sisters gave it all that day, as did the brothers and sisters who worked at Ground Zero in the days and weeks that followed, and those of us who were able to make a difference with Katrina. I'm not saying that the events of 9/11, happened for a reason; what I'm saying is that those we lost on that day continue to help us. They helped save the lives of tens of thousands of people on the Gulf Coast and they weren't even there. They made it possible for us to think out of the box, plan for future disasters, do what we do better, and help the people who need us most. We did lose them and it hurts, but they are helping us more than ever right now. They will as long as we're willing to let them help. Their loss has helped us to realize several things. We have to make a difference while we can. We have to build our own legacy, mentor our people, invoke change, take better care of each other, train better, and fight for better equipment and staffing. We must remember who's at home waiting, and that each and every day, hour, minute, and second is a gift and that we should not waste it by hurting each other or by being in a bad mood.

The hugs feel better than ever. You hear firefighters calling each other brother and sister more. Keep it up or it will be gone before you know it. After 9/11 Bill Manning wrote an editorial called "They are alive in you." They are. Keep them there. Remember what they did for us and what every brother and sister who has gone before has done for us. We owe it to them and to all of us as well. *Never forgetting means never forgetting.*

With all the bad things that happened on 9/11, it's hard to see the good that came from it. Don't let our brothers and sisters go in vain. Remember the "Stop sign." Make a difference, seize the moment. We are a little bit better.

9
CEREMONIES THAT STOKE THE FLAMES OF TRADITION

Our fire service is rich in tradition and has a history so fascinating that as a firefighter, young or old, the more you dig into it, the more you want to learn about how it all started in this country and about all of the changes in direction the service has taken over the years to bring us to where we are right now. Many of the changes have been good and have brought us to new levels, but a few have cut away at the very foundation that supports the fire service. Change can be good, but a lot depends on how you go about making changes. It's hard to watch some in our profession working so hard to take the firefighter out of the firefighter and the firehouse out of the firehouse when we should be trying to put them back in. There's nothing wrong with a firehouse looking like a firehouse instead of one of those in-and-out, lube- and oil-change places. There's also nothing wrong with letting our firefighters dress and act like firefighters.

There Is a Business Side to What We Do, But...

So many seem to be trying to turn us into a business and hide us from the public. There definitely is a business side to what we do. But if you want to know why you can't get your guys to do something for you or can't understand why they seem to have lost their love for the job, think in terms of traditions, not business. Maybe bringing back some of the traditions that we used to have is the way to stoke those fires your guys have within them. Maybe that's all you need to do to get them acting and looking like firefighters. Maybe, just maybe, you could rekindle their love for the job. Trust them. They're extremely smart, talented, and professional, and they will do a great job if you let them love the job just a little.

The right tradition is not a bad thing

The first thing you hear from some people when you mention the word tradition is that the only thing tradition does is keep us down. They say that tradition opposes change, that it's stubborn and closed to new ideas. They say that it's tradition that gets us hurt and killed doing the same wrong things over and over again. When it comes to hurting and killing our own, *we do need to change the things that get us into trouble*. We should try to implement anything that will keep our firefighters healthy and safe.

But when I speak of tradition I'm not referring to the habits that have hurt us in the past. I'm talking about the kind of tradition that celebrates our heritage and what we're all about, the traditions that come with the greatest profession in the world. There are a lot of great and wonderful things that have been part of the fire service for decades and we should reach out to them, because if we're not careful, they'll keep slipping away one by one until they're all gone. That kind of tradition is good. We often refer to our department as an extremely traditional fire department that is very progressive. You *can* have it both ways!

Ceremonies

As you begin your trip back in time, looking for and learning about our fire service history, what you start to see is that there have always been a variety of ceremonies and celebrations. There are so many that it would be difficult to describe them all in this book. But a lot of people have asked about them so that they can bring that type of tradition in or back to their department, so let's take a look at a handful of them. Let's look at some of the ceremonies that promote pride and ownership in the fire service and provide an avenue for developing that love for the job.

The new firefighter

The first and perhaps the most important ceremony is the swearing in of a new firefighter. This ceremony is where it all starts. It is our opportunity to make an impression, to have an impact, and to set the tempo for what's coming up in the firefighter's new career. Some departments hold this ceremony when they hire a new member and some wait until they have completed and are graduating from the Fire Academy. Either way, there is a reward for performing this ceremony.

First, we're confirming to the new firefighters that they are about to become part of a very special family that is second only to their families at home. Second, they know right from the beginning that they're going to be held accountable. At this ceremony we let them know that it is a privilege and an honor to wear our badge. Make them raise their hand and swear to do the right thing (fig. 9–1).

Fig. 9–1. This is where we let them know it's an honor to wear our uniform.

They should swear to honor those around them and to carry on our tradition. Too many departments just hire people, tell them to report for duty, and later wonder why their firefighters don't seem to care about or appreciate their job and the department.

This is also a time to celebrate, for them and us. For them because they are entering the most fascinating and most rewarding job anywhere and for us because we're adding to our family. It's a time for their families at home to be proud as well. This is how our department handles this one; obviously you can add to it whatever you feel will make it more special:

1) We set a date and time for the ceremony that works for both the new hires and the staff.

2) We hold it at our Firefighter Memorial (weather permitting). It provides a very special backdrop and serves as a reminder of those who have sacrificed and gone before us and of what we owe them (fig. 9–2, 9–3).

3) As many stations attend as possible and staff members wear Class A uniforms.

4) The new members' family and friends are invited and the family members are asked to stand behind and just to the right of the member being sworn in. The members then introduce their family and guests (fig. 9–4). This is a big day for them as well and it's nice to have them share in the moment and participate. Also at this point, one of the assistant chiefs introduces any dignitaries in attendance (fig. 9–5).

5) The ceremony starts with an introduction and welcome by the fire chief. When possible, our chaplain provides an invocation (fig. 9–6, 9–7).

6) Next, the City Secretary has the new members raise their right hand and take the oath of office (fig. 9–8). To make it easier on the nervous members, she provides the oath in writing for them to hold onto and read along.

7) After the new members have taken the oath of office the mayor says a few words and then pins their badges on them (fig. 9–9). What can be a nice touch at this point is allowing a family member who is a current or retired firefighter to pin them. When we can do that, we have the mayor shake the member's hand and give the badge to the person that is going to pin them. It's a really nice touch and brings back that family feeling again.

8) We also provide a printed program of the event and take lots of pictures.

Fig. 9–2. Our firefighter memorial serves as the perfect backdrop for the ceremony.

Fig. 9–3. Lewisville's memorial includes a portion of a box beam from the south tower of the World Trade Center; this adds meaning to the memorial and ceremony and honors those who have sacrificed.

Fig. 9–4. This is a big day for the members and for their families.

Fig. 9–5. The ceremony is attended by family members, citizens, and dignitaries.

Fig. 9–6. The fire chief welcomes those in attendance before introducing the mayor and city secretary.

Fig. 9–7. No ceremony should proceed without either a moment of silence or an invocation from your department chaplain. Lewisville Fire Department Chaplain Benny Grissom is always on the agenda at our ceremonies.

Fig. 9–8. The city secretary provides each member being sworn in with a copy of the wording to help combat the nervousness that can sometimes creep in.

Fig. 9–9. It's a perfect time to have the mayor say a few words about the fire department and the members being sworn in.

The promotion

At the promotion for one of our members we follow the same format used to swear in new members but we add one very special step. After the mayor or family member pins on the member's badge, we conduct a "collar pinning." We allow the member being promoted to invite one or more special people to assist them by pinning their collar insignias (fig. 9–10, 9–11, 9–12). The person they choose could be their spouse, their children, a family member, a friend, or a mentor.

Fig. 9–10. Having a member of their family pin their collar pins involves the special people in their lives in the ceremony.

Fig. 9–11. The collar pinning should be done by a person that is in some way special to the member being promoted.

Fig. 9–12. Having the mayor pin a new member's badge can add importance to the occasion and to the member's new position.

It's a big day for their family as well

It's a big day for family as well as for the member being promoted. They have been at the member's side, putting up with the longs hours of studying, mood swings, the up times of making the list and the down times of not coming out on top, and have supported them throughout the whole process. *It is a big day for them, as well, and an extremely proud moment.* At one promotion ceremony, the member being promoted to captain had his long-time captain pin one of his collar pins. When it was over there wasn't a dry eye in the house. It is really a special moment and adds class to the event. And when the ceremony is concluded just as it is with the new firefighter swearing in, the department provides cake and coffee and a reception area for those that attended to gather, visit, take pictures, and offer congratulations.

Retirements

This is a ceremony that has to be special. When one of our members gets to the point that they can retire, it should be celebrated and they should be congratulated. To wish them well is the right thing to do. Now, I know that there are many firefighters who say that when the time comes they want to slip out like they slipped in. They don't want parties, tributes, or speeches; they just want to leave quietly. I can respect that, but I always warn them that they're going to regret it down the road. They're going to miss saying good-bye.

And, often, failing to honor a retiree is rough on the troops as well. When they don't get to say good-bye or wish the retiring member well, it leaves them with this hole, or void. So many departments don't do anything for a retiring member and that's a shame. Considering all that our members put into the department during their careers, all the lives they affect, and the impact they have on their brothers and sisters, it's only right to honor and thank them. We owe it to them!

When one of our captains, Butch Flanagan, decided to retire after serving 35 years, I remember him explaining what happened his last shift day. He said that it was about 6:45 AM and he was about to end his shift for the last time when a call came in. The oncoming shift said they would take it and off they went. He said he sat there for a couple of minutes and then looked at one of his firefighters and said, "I guess this is it. It's over." He then loaded up his pickup truck and headed home. Can you imagine if we hadn't had something planned for him? We did, the following shift day, but can you imagine allowing a guy like that, one of the most respected, well-liked, and best captains I've had the pleasure to be around, just walk out the door after 35 years of service with no party, no retirement ceremony, nothing? That's just wrong, period. If you allow that to happen in your department, change it! Do something special for your retiring members. If you don't think it's necessary or just plain don't want to, shame on you! The following is how we try to do it in Lewisville.

1) Plan to hold the ceremony on their last day and at headquarters.

2) Take their company out of service during the ceremony and try to bring in as many stations and companies as possible. The retiring member and staff wear their Class A uniforms.

3) The fire chief starts the ceremony with an introduction and welcome and the chaplain provides the invocation.

4) If the member is an officer, it is at this point that all members of their shift, company, or, in the case of a chief officer, as many from the department as possible, line up and come to attention (all wearing their Class A uniforms). The retiring officer, flanked by the chief of the department, conducts a final walk-by and inspection (fig. 9–13, 9–14, 9–15, 9–16, 9–17). When he or she reaches the last firefighter, the order is given to present arms and all members salute. The retiring member returns the salute, the members order arms, and the honor guard or a chief officer presents the retiring member with a flag that was flown over their station on their last day. An additional nice touch here is to have your pipe and drum unit play a tribute song.

5) At this point the station paging tones for the retiring member's company or the all-call tone for a chief are activated and a dispatcher announces the member's retirement, thanks him or her, and wishes him or her well (fig. 9–18).

6) At this time everyone is dismissed and asked to gather in an area or room where presentations can be made. This is the time that gifts and tributes are given and finishes with everyone getting that chance to say thanks and wish them well. It's the perfect opportunity for some good stories as well (fig. 9–19, 9–20, 9–21).

Fig. 9–13. As many department members as possible line-up for the final inspection and line up by shift assignment.

Fig. 9–14. The retiring member is flanked by the chief of department as he begins his final inspection.

Fig. 9–15. Next, the retiring member walks down each line inspecting the personnel.

Fig. 9–16. The honor guard presents the member with a flag that was flown over their firehouse on their last day.

Fig. 9–17. Just as important as the chaplain to a ceremony, the honor guard adds the presence of honor and selflessness.

Fig. 9–18. Probably the most emotional part of the ceremony, broadcasting the last alarm for the member over the radio.

Fig. 9–19. After the final inspection, flag presentation, and last alarm tone, everyone moves inside where presentations are made and accolades are shared.

Fig. 9–20. Chaplain Grissom presents Captain Flanagan with his helmet to keep as a memory and keepsake.

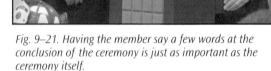

Fig. 9–21. Having the member say a few words at the conclusion of the ceremony is just as important as the ceremony itself.

Graduations

This ceremony is fairly easy. Give your fire academy or community college a call and ask them for their line-up. They've been doing graduation ceremonies for a long time, and based on their ideas, you should be able to put something really nice together. One key point here is to get a good keynote speaker. There are many talented people out there who can get your graduates fired up and have them sitting on the edge of their seats. This is a great way to send them out with a bang. It's also a good time to remind them about what the fire services is all about. And don't forget to have your honor guard start things off by presenting the colors.

Awards ceremonies

Whether you hold it as a banquet or in the firehouse, an awards ceremony is very special. One of the hardest things to do in our business is recognize a firefighter for a job well done. The immediate response from most firefighters is always, "No thanks, Chief. I was only doing my job."

And that's not bad, but we still need to recognize them. I was privileged to have Marine Corps Major Jason Frei speak at our Heroes of Denton County banquet after he presented at FDIC in Indianapolis. Major Frei lost his right arm in Operation Iraqi Freedom in Iraq and had a wonderful and inspiring message to share. His speech was great, but there was one thing that jumped out at me. It was the part about receiving medals and wearing your uniform. He said that he wears his uniform out of pride and to honor the Marine Corps. He said that he also wears his medals for the same reason. Not out of arrogance but out of pride and honor.

Wear them to honor those around us and our service

When Major Frei was done, I realized what he was trying to say. The medals that are given to a firefighter for an act of bravery or for an act needing a commendation are not really about the recipient. They're about what we stand for, and they honor those who went before us and those we will lose in the future. While it is true that the medal is being given to you for something you did individually or as part of a team, it's pretty selfish to think it's all about you. Major Frei said that you should wear your medals proudly, and when somebody asks you what they're for, explain it to them with the pride you feel for the fire service, an organization that is one of the best in the country. Accept your medals on behalf of the fire service and your department and accept them with honor.

I know that it's our job to do the things we do, but when an incident or action is above and beyond or has impacted someone in a special way, we need to recognize that kind of effort. If we don't, no one else is going to. Below is a description of how we handle our awards ceremony.

1) The awards committee plans the event, makes the award selections, and really puts in a lot of hard work getting the ceremony put together. There are many variations on the award ceremony process and most are very good. You won't have to look far to find a good one. We hold a banquet, but having the ceremony at a firehouse with refreshments also works well.

2) A date is set, a location determined, and we ask for station coverage from our neighbors. Those members on duty still staff their companies but because of the station coverage by other departments they don't have to push out on every call. If needed, they can respond, but this arrangement allows

them to join in with the festivities. This is not a problem for on-duty members because alcohol is not served at our ceremony.

3) All members wear their Class A uniforms.

4) As is almost always the case, the honor guard starts things off by presenting the colors. After everyone is seated, the fire chief welcomes those attending, one of our assistant chiefs introduces any of our special guests (the mayor, council members, visiting chiefs, retirees, etc.) and our chaplain provides an invocation.

5) At this point, dinner is served. As soon as dinner is over a keynote speaker addresses the group.

6) After the keynote speaker concludes, we show a musical tribute video highlighting our personnel. The footage includes the past year's calls, events, and activities. Each member is presented a copy of the video as a gift. One benefit of this is that each member will have a pretty good documentary of their department and their career, something we had never really done before.

7) Next is the presentation of the awards. We start with years of service, then honorary mentions, awards of exemplary action (our award for civilians, police officers, and so on), company citations, commendations, awards of merit, the medal of valor, and finally, our rookie, firefighter, paramedic, and officer of the year awards. One nice touch: every now and then we invite a victim—usually a CPR save—to the banquet. We don't tell anyone that they are coming and when it comes time to recognize that company who saved the victim we have this person join us on stage and help present the awards. The victim really enjoys it because they get to say thank you one more time and it demonstrates to all in attendance just how special our people are and what kind of an impact they can have on someone's life.

8) Then our master of ceremonies wraps things up, thanks the committee for all of their hard work, and wishes everyone well. Some other nice touches include asking whether any of the spouses would like to keep the table centerpieces,

having a photographer on site to take photos of the troops, companies, families, etc., and just making sure everyone feels comfortable and at home. It's a great night for our families.

New apparatus

As you dig deeper into our history, you discover that we have always had ceremonies when placing a new rig into service. Ceremonies have ranged from giving the rig its first bath or a wetting down to actually pushing the rig into the firehouse like in "the olden days." But it seems that over the years we have let this tradition go in many departments, just like we stopped taking pictures of the rigs, or, worse yet, continued to take pictures of the rigs but forgot to put the guys in the pictures (fig. 9–22). This ceremony in particular really fosters pride, and when done well, can help some firefighters understand why we take care of our rigs in the first place.

Fig. 9–22. All four of these new ambulances were presented and placed in service on the same day as part of our new apparatus ceremony.

Our new apparatus ceremonies look like this:

1) First we send out notice of the event with the date, time, and location.

2) Begin by having the chief welcome everyone, introduce any special guests, and have your chaplain provide the invocation.

3) At this point you can "wet down" the rig before pushing it in. The company officer usually handles this.

4) Next, have your dispatch office activate that station's paging tones, announce the retirement of the old rig, and welcome the new one aboard.

5) Then comes the fun part: pushing the rig into quarters. One note, as heavy as rigs are today, it works well to have a driver in the rig backing it in slowly while all participating simulate the push. This is pretty cool for the troops as well as for your citizens if you want them to participate. Then it's cake and coffee time.

Our fireboat

When we dedicated our fireboat, we modified the new apparatus ceremony. We did steps number 1, 2, and 4 from the list above, had our chaplain bless the fleet, and, instead of the "push," we announced and revealed the name given to the boat. It's always been said that it's bad luck not to name a boat. Next we christened the boat with a bottle of champagne (fig. 9–23, 9–24, 9–25, 9–26, 9–27). The christening was done by the member after whom we named the boat. By the way, we christened two boats and neither of the members after whom the boats were being named knew what the names were going to be until they were revealed during the ceremony. It was a very special moment for both of them and for our fire department.

Fig. 9–23. Placing a new fireboat in service is important, and the ceremony is supposed to bring good luck to the unit and its crew.

Fig. 9–24. The "blessing of the fleet" is the first official act performed during this ceremony.

Fig. 9–25. Boat 168's namesake, Bryan Jarvis, had the honor of christening the boat with a bottle of champagne.

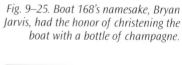

Fig. 9–26. The name of the boat is displayed in a prominent spot for all to see.

Fig. 9–27. Pictured next to the boat named after him, retired Division Chief Michael Paul Alaga takes time for a proud moment.

New firehouse

A new firehouse dedication is a special event for more than just the department. It's special for the neighborhood the firehouse is going to serve. When we dedicated our last firehouse we also held a mini fire prevention week open house with a variety of activities and events for the families and businesses that surround the firehouse. Following is a description of that ceremony (fig. 9–28).

1) As in all of the ceremonies, the fire chief gave a welcome and introductions were made. A short speech was made to acknowledge what the opening of the firehouse would do for the community.

2) All of our special guests were asked to line up in front of the bay doors for the ribbon cutting, but instead of cutting a ribbon, we conducted a "hose uncoupling." The mayor and fire chief stood across from one another at the coupling and the mayor was asked to uncouple the hose while the rest of the guests held the hose at waist level, creating a look similar to the ribbon that you usually see at a dedication like this.

3) As soon as the mayor began to uncouple the hose, the firehouse alert tones for the new firehouse were set off and an announcement was read that welcomed the firehouse and its personnel aboard and wished them a safe journey. Then we began the open house.

Fig. 9–28. Rather than the normal ribbon cutting at the opening of a new firehouse, a "hose uncoupling" adds the fire service touch to the ceremony.

The 20-Year Anniversary Firehouse Dinner

Another ceremony that comes to mind is the 20-Year Anniversary Firehouse Dinner. One of my battalion chiefs, Jerry Wells, was visiting in New York. While riding out with a battalion chief, he was told that they were going to be stopping by a firehouse where the guys were celebrating the 20-year city anniversary of a couple of their members. The FDNY battalion chief explained that they celebrated by preparing a nice meal and inviting the members' families to the firehouse, making it more than just another day at the firehouse. It may not seem like a big deal—just dinner at the firehouse—but it was important to the two guys celebrating 20 years and to those who got to prepare the meal for them and their families. When my battalion chief returned from his visit, in addition to all of the other things he learned, he decided to hold an anniversary celebration for the members on his shift who were about to celebrate their 20-year department anniversary. He even invited a couple of the guys from the same academy class who had started on the same day. Now he holds a celebration for everybody on their 20-year anniversary, even if it's just cake and coffee.

Be Careful, Because When They're Gone, They're Gone

It's through these types of events that we can continue to strengthen our foundation. If we don't, I guarantee you that these traditions will all disappear. And once they're gone, they're gone and are hard to bring back. We've gotten to the point that the only time we have a ceremony is when we lose a firefighter and the only time we wear our Class A uniforms is for a funeral. That doesn't have to be (fig. 9–29). I wear mine when I am at the majority of my public speaking engagements in town or when I'm representing my department because I'm proud of what I do.

Fig. 9–29. A funeral or memorial should not be the only times we wear our Class A uniforms. They should be worn proudly any time we represent our department at a formal function.

The Time to Honor Someone Is When They're Still with Us

We got a few things mixed up a while back. The time to honor someone is not when we've lost them; that's when we pay tribute to them. The time to honor them is when they're still with us, next to us, right in front of us. Let's not wait until they're gone; let's do it now. So celebrate, hold ceremonies for all of your special events, and steer your firefighters in a direction that will rekindle their love for the job.

These are just a few, very few, of the ceremonies that have been conducted throughout the years and there are many variations on each one. Whether it's one of the ceremonies described above, a firefighter memorial dedication, one in which you make someone an honorary chief or firefighter, or any others you want to conduct, they're all important (fig. 9–30, 9–31, 9–32, 9–33). In my eyes, there is no higher honor for a citizen than to be made an honorary firefighter or fire chief. Look for opportunities to have ceremonies. I've been blessed with many friends who have shared their ceremony ideas with me and that's something that we all should be doing.

"The time to honor them is when they're still with us ... Let's not wait until they're gone; let's do it now."

Examples of the various program flyers, announcements, and forms that were mentioned in this chapter are provided in the back of this book or you can give me a call. That way you can maybe use some of it for your own ceremonies. Hey, that's what we did!

Fig. 9–30. Honorary chiefs: Dallas Stars hockey trainer Dave Surprenant and goalie Marty Turco.

Fig. 9–31. In addition to being made honorary chiefs, actor John Travolta was made honorary mayor and actor Joaquin Phoenix was given the key to the city.

Fig. 9–32. One of the highest honors that can be bestowed on any civilian is to be made part of our family.

Fig. 9–33. During the ceremony, time was set aside for a tour of the firefighter's memorial.

10
MARKETING YOUR FIRE DEPARTMENT

Whenever someone mentions customer service or marketing in the firehouse, a lot of our firefighters and officers lean back in their chairs, wrinkle their noses and say, "Not again! What are they going to make us do now?" It wasn't too long ago that I felt the same way. The idea of customer service being pushed and shoved onto us was not something I thought I needed to be interested in. It had no part in what I thought of as a firefighter's job. I didn't remember my dad doing it years ago when he was a firefighter in a southwestern suburb of Chicago. I remember him cutting a lot of holes in roofs and busting a lot of windows, but I don't remember him talking about customer service.

So that was it. My mind was made up. I was going to do the three things that one of my fire service idols, Chief Jack "Mac" MacCastland, said were the best things about being a firefighter: you can break things, swear, and get dirty. All three things your mom never wanted you to do. Now don't get me wrong, Mac is one of the most forward-thinking people I know, but it was kind of cool to think about firefighting that way. So I wasn't the least bit interested in this new customer service thing. Well, that was when I was younger and needed a lot of guidance and the occasional whack on the head.

We Need To Market What We Do, All of It!

But with all that has happened since September 11, 2001, we've realized that if we don't do a good job marketing our fire departments and our profession, we may be squeezed out of the picture entirely. Just take a look at what's going on in some cities where the police department is taking over EMS, haz-mat, arson, homeland security, emergency management and the list goes on, leaving some in the fire service feeling helpless as they see their world self destruct and disappear. Well, we don't have to sit by and let it happen. For years, police departments have been using creative techniques to market themselves and get the things they want. I hate to say this, but the fire service could learn from them how to market *our* mission.

Marketing Our Mission

A few years ago, Matt Mosley and David Rhodes, both with the Atlanta, Georgia, fire department, gave a presentation at FDIC in Indianapolis called "Marketing the Mission." Matt had been involved in a helicopter rescue that received national live media coverage and he and Union President Rhodes used it to their advantage to get things for their department that they desperately needed. The point of their program was to explain their success but also to demonstrate how you could accomplish some pretty neat things just by marketing what we do. Matt and David suggested going out and letting people know what we're all about and that we're not just sitting around and waiting for the bell to go off. They told us to be honest and tell the public that we need some things, that some of our fire departments are hurting, and that we're trying to do our jobs with little or no funding while other city departments get whatever they need. The public will realize that something is wrong when they see the amount of money going to law enforcement while fire departments are fighting to keep personnel and holding fundraisers for thermal imaging cameras and other equipment.

It's Not Really New

What I've come to realize is that customer service *is* part of a firefighter's job. When I thought about it, I realized that we were doing it already and had been for a long, long time. In fact, my dad *was* doing it back in the sixties. They didn't call it customer service back then; they called it boarding up the hole in the roof, covering the windows, cleaning up a little, and taking care of the family after the fire. I was under the impression that in order to "do" customer service I had to give up throwing water, cutting holes, pulling ceiling, and all of the other stuff. I had that very conversation with Phoenix Fire Chief Alan Brunacini and he said, "You know, you can still do all of that. We just need to help out the folks when we're done (fig. 10–1)." This coming from the mouth of the Godfather of customer service in the fire service.

Fig. 10–1. Chief Alan Brunacini

Let Us Be the Shining Star for a While

So we're doing it in Lewisville, Texas, and we did it in Coeur d'Alene, Idaho, and the boys up north are still doing it. It was easy. We told the guys, "The more stuff you do, the more stuff you get." (This obviously depends on your city's economic situation. But what do most city councils, mayors, and city managers want, problems or good things? Enough said.) Why not go out and wow the community instead of letting the cops beat us to it? Let us be the favorite son or the shining star for a while, we told our guys. What's funny about it is the troops love doing customer service in this way. They're having fun. They love the gratitude and the community's new-found appreciation for the fire service. They feel great because they're helping people and that's why most of them got into this business in the first place.

This stuff really works

There are a lot of activities that the fire service can be doing to market our mission. Believe it or not, many fire departments and firefighters are already doing them. They're just not telling the public about them. Fire departments are already assisting families and businesses after a fire, reading in their schools, providing public education programs, and helping those in need. They just don't go out and brag about it. Well, it's time to *start* boasting about *all* we do. There's nothing wrong with explaining to the public that we do more than put out fires and cut people out of cars and all of the news-worthy stuff. We need to let the public know that we are truly an asset to the community and that we offer much more than other public service agencies. I don't mean to knock them, but it's true. I know it's hard to boast because firefighters are modest, hard-working people who do what they do because they love it. They don't do it for the money, we all know that, or for the glory, but because they love helping people. But we need to get past that modesty, set aside our pride, and go out and do some bragging and fill people in on all the different things we do. Firefighters are a very proud breed, proud of all that they do.

I've always found it amazing and kind of fun to give a talk to a service organization or homeowners group, and see the surprised look on their faces when I tell them all that our firefighters do. Time and time again, they come up to me afterwards and say that they really didn't know that we did all of that. And often they ask what they can do to help. Our department does very well each year budget-wise, and I truly believe that it's because of all that our firefighters do and because we make the effort to go out and tell people that we're doing it!

Some Successful Programs

"Vested for Life" program

Lewisville is located just north of Dallas and has a recreational lake, Lewisville Lake, that sees over 2.7 million visitors annually. More than 300,000 residents live adjacent to Lewisville Lake or have indirect access to it. It's a little more than 35,000 acres in size and is one of the busiest lakes in Texas. It is a pretty busy place for us, generating more than 250 incidents each year. We see a lot of boating accidents, stranded boats, and a variety of other calls, but we also see as many as 5 to 13 drowning incidents a year. When we investigate the circumstances of each drowning incident, one thing becomes clear: we've never pulled anybody off the bottom of the lake who was wearing a life jacket.

So, with that in mind, we implemented a program called Vested for Life. On weekends between Memorial Day and Labor Day our dive team goes out to the beaches and swimming areas and gives away free swimmer's life vests to any child who doesn't have one (fig. 10–2, 10–3, 10–4). We leave a lake safety brochure written in both English and Spanish with the parent or guardian and go on our way. We're hoping that we can bring down the number of drowning incidents and near-drowning incidents.

Fig. 10–2. We've never pulled anyone off of the bottom of the lake wearing a life jacket.

Fig. 10–3. Dive team members realize a huge level of satisfaction in distributing life vests, knowing that this act will help to save lives.

Fig. 10–4. In a matter of minutes, the dive team can distribute life jackets to all of the kids at a particular swim area or beach.

"Blazing a Trail for Literacy" program

Another program that is impressing the public is one of many that we do for our elementary schools. We call this one Blazing a Trail for Literacy. It's a program that encourages children to read. Reading is very important and is a fundamental building block in the development of our children. We try to promote reading in every possible way because with good fundamentals children grow into good contributing citizens. As fire chief, I read to children at the library a couple of times a month and I try to get to the elementary schools as well.

Marty Turco, goalie for the Dallas Stars hockey team and Lewisville Honorary Fire Chief, kicks off the program by being a guest reader at some of our schools. The kids love it and he loves it as well. It fits right into his own public service program, "Stick with Reading." Our reading contest involves the kindergarten through fifth grade classes at all ten of our elementary schools (fig. 10–5, 10–6, 10–7, 10–8). At each school, the student from each grade level who reads the most minutes is declared the winner. The six students from each school are rewarded for their hard work with a backpack with our department patch on it, a gift card to the local book store, an autographed hockey card or hockey puck from goalie Marty Turco, and a ride to school on a fire truck, which arrives at their school in front of all of their friends. The kids love it, the parents love it, and the teachers love it. If I had known that offering a ride to school in a fire truck was going to spark as much interest in reading as it does, I would have done it years ago.

Fig. 10–5. This photo is used to promote the fire department reading program.

Fig. 10–6. Honorary Chief, Dallas Stars hockey goalie Marty Turco reads to a third grade class as a guest reader for the fire department.

Fig. 10–7. Having a guest read once in a while—especially one like Marty Turco—helps keep the kids glued to the story.

Fig. 10–8. The winners of the reading contest get a ride to school on a fire engine as part of the grand prize.

"After the Fire" program

One of our biggest programs is called After the Fire. What's great about this program is that a lot of fire departments are already doing it and more and more are starting the same kind of program. We wanted to take it to the highest level that we could. Rather than handing the victims of a fire a blanket when we're done and wishing them well, we wanted to make sure that we did whatever we could to get them through a very hard and sometimes heartbreaking time.

The program is broken down into two areas: "After the Fire Residential" and "After the Fire Commercial." The residential part of the program is set up to assist residents who have suffered a fire, a flooding incident, or anything that results in damage to their home or displaces them from their residence.

We have used the Red Cross for years and we continue to do so, especially with a major loss or one involving a multiple-family dwelling, and they do an awesome job. We just serve as the launching pad for the kind of assistance the Red Cross can provide. So, with that in mind, we are able to provide the following, often while we are still fighting the fire.

First, a representative from our department is assigned to assist the residents (fig. 10–9).

Our representative helps determine the needs as to how we can help the residents, which, depending upon the severity of the incident, may

Fig. 10–9. A fire department representative meets with the family as soon as possible to help determine their needs.

be anything from giving them our "After the Fire" booklet, which contains suggestions to get them back on track, to providing any or all of the following services:

1) Contacting their insurance agent and getting an adjuster out quickly. Sometimes a call from the fire department gets a faster response.

2) Attempt to have them identify any specific or special items in the home so that we can try to retrieve them. It's understood that everything in the home is important and special, but often there are specific items such as paperwork or family keepsakes that they really need and are worried about.

3) Determine whether they have a place to stay. Several of our local hotels offer a free stay to those who have suffered a loss. This can be anywhere from one to three days or until the insurance company takes over. The hotels will work with those who don't have insurance as well.

4) Several of our local restaurants have provided coupons for meals so that we can ensure that those involved in the incident get fed.

5) If they are in need of medication that has been destroyed or is inaccessible, all they have to do is have their doctor contact one of our pharmacies. The pharmacy will fill the prescription for free, anytime, 24 hours a day.

6) If they need eyeglasses, hearing aides, clothing, toys or furniture, we can get them for them.

7) We have a public storage facility that will store their belongings for four months for free.

8) We have boxes, bags, and storage bins for their belongings.

9) If they need toiletry items such as toothpaste, a toothbrush, deodorant, and so on, we have it for them, including diapers and wipes for infants and toddlers.

10) If they have pets we can get them free emergency care or overnight boarding if needed.

We have so much to offer those in need that the list goes on even further from here.

Our customer support unit

Recently, we placed our Customer Support Unit in service. We took a two-ton ambulance (which had been replaced by a new front line rig) and converted it into a rolling toolbox. It carries just about everything that you can think of to assist someone after a fire: plywood, lumber, plastic, salvage covers, water vacuums, shop vacuums, generators, tools, and a very long list of other items our troops thought necessary to assist those suffering a loss (fig. 10–10, 10–11, 10–12, 10–13). We've used it several times already and each time it receives compliments from those we help. And, for the most part, everything for this program, including the items carried on the customer support rig, has been donated. The local lumber and tool supply stores actually tried to outdo one another, competing to see who could donate more.

Fig. 10–10. This unit is designed to assist families and businesses after the fire.

Fig. 10–11. This unit is popular with the public, but the mayor and council love it as well, which brings its own set of benefits.

Fig. 10–12. Everything you need to take care of a post-fire situation is carried on the unit, including tools and lumber.

Fig. 10–13. Within a matter of minutes a hole can be covered, helping to reduce further damage

And as for a commercial fire, we can move a business to another location quickly and get their phones transferred so that when one of their customers calls on them the next day, someone will answer. We can offer them free office supplies and office furniture, including free computers, printers, and fax machines. Most of our businesses are not large corporations, though we treat them as if they were. But large corporations usually have a risk management and recovery team to take care of the situation after a fire. So it's the majority of businesses out there, the smaller ones, for whom we need to provide the most help. In some cases these small businesses are hanging on by a thread; when a fire hits them, it could mean the end of their business as they know it. Even long-time customers may have to do business elsewhere for a while and there's no guarantee that they will come back. That's one reason why many businesses fold about a year after a fire loss. But we work with them for days or even weeks to ensure that they make it back to some kind of normalcy. Every business we lose means a loss in revenue to the city, the loss of jobs, and another vacant property.

Providing the foundation for stability

We want to provide the foundation for stability in a time of crisis and the businesses that donate to our cause love doing so. They get a chance to help someone in need and they know it's good for business. If I suffer a fire loss and the Residence Inn opens their doors to me and my family during such a traumatic time in our lives, you can bet that I'll stay in one of their hotels when I'm traveling, because they treat people like family.

"Santa Claus" program

Another program that gets a lot of praise and is loved by the public is our Santa Claus program. We give Santa a ride on the engine or truck and tour each neighborhood with our lights and sirens going, handing out candy (fig. 10–14). The kids love it, the parents love it, and, best of all, our bosses love it. For the most part, the troops love it, too.

Fig. 10–14. Our Santa program is regarded by the city council as one of the best public relations programs the city has ever done.

"Opening Day at School" program

Our most recent program deals with our school zones. Every year, during the first week of school, we send a piece of apparatus to each elementary school at morning drop-off time and at afternoon pick-up. We park the rig in a highly-visible spot, turn on the warning lights, and place a banner on the rig that says "Please drive slow, school is back in session." And it's working (fig. 10–15, 10–16, 10–17). The cops are even saying, jokingly, that our program is "muddying up" their fishing hole because drivers are slowing down and they're not writing as many tickets. It's an awesome reminder for the drivers passing through the school zones that school is back in session and it is our hope that fewer children will be hit by a car.

Fig. 10–15. Each piece of apparatus is parked in a highly-visible position with red lights flashing in an effort to slow drivers down on the first three days of school.

Fig. 10–16. The banner helps explain what the message is for the drivers.

Fig. 10–17. The police, happily, have fewer tickets to write because drivers are slowing down in the school zones.

And it doesn't stop there (fig. 10–18, 10–19, 10–20). As at so many other fire departments, the troops give away free smoke detectors and free batteries, hold an annual fire prevention week open house and a mini fire prevention week open house for the foster children of Denton County, and contribute to and assist with public education programs throughout the year.

Fig. 10–18. Kids are given not only a good view but hands-on access to the fire equipment.

Fig. 10–19. Being able to spray water at a simulated house fire with a real fire hose is a big attraction.

Fig. 10–20. Open houses are a great way for community organizations to reach out to the public.

Following is a short list of some of the other programs that we do that get us great results and help market our mission

- Media Fire Academy, where members of the media are provided with a look at what our fire department does. The day starts with a review of the history of the fire service. Then they get geared up in turnouts, are assigned to a piece of apparatus and head out to Lake Lewisville, where the dive team gives them a demonstration. Next they're off to the auto salvage yard to work with the extrication tools opening up cars and packaging simulated patients. Then it's lunch at one of the firehouses and out to the training field, where they are taught how to handle hose lines, wear an SCBA, cut with the saws, and perform a search. To top it off, they end up fighting fires in our burn tower. They attack the fire, search with a thermal imaging camera and without one, cut burglar bars, cut holes in the roof, and get dirty (fig. 10–21, 10–22). At the end, we hold a "field" graduation ceremony where they are given a diploma, a Media Fire Academy t-shirt, a hat, and a reflective vest that says "Media Team" on it to wear at incidents.

Fig. 10–21. Providing a reporter with a front row seat in a live burn situation allows them to see what we see inside a burning building and they are always impressed.

Fig. 10–22. The trust and the relationship developed during the academy carries onto the fireground and makes for better relations with the media.

- Citizen's Fire Academy, which is a great program and a great public relations tool (fig. 10–23, 10–24).
- Fire Explorer Post 911 is a great recruiting tool and serves as a very nice cadet program (fig. 10–25).

Fig. 10–23. Firefighter Wayne Davis instructs a citizen in the proper way to check for a good seal with their facepiece.

Fig. 10–24. All of the citizen fire academy graduates are given a class picture as a keepsake.

Fig. 10–25. One of our Fire Explorer members, Whitney Weber, assists during open house as well as with other functions throughout the year.

- Our "Public Education Clown" program visits every elementary school in the city and performs at several public education functions (fig. 10–26, 10–27, 10–28).

Fig. 10–26. Ashes the clown always has a great way with the kids.

Fig. 10–27. Radio the clown uses humor as a way of "sneaking up" on the children with fire prevention tips.

Fig. 10–28. Fire prevention "clowning" has been a great way for the fire service to get out fire prevention messages for years.

- "Shattered Dreams" is a collaborative program with our police force and fire department dealing with teens drinking and driving (fig. 10–29, 10–30, 10–31, 10–32, 10–33).

Fig. 10–29. Using the same materials used in a disaster drill adds realism to help make a point.

Fig. 10–30. Using several victims from their own class helps to bring the potential for tragedy close to home.

Fig. 10–31. Being able to actually cut a car apart adds realism.

Fig. 10–32. The simulated arrest drives home the point of managing the consequences of your decisions.

Fig. 10–33. At the conclusion of the program, a message is delivered to the student body by a member of the fire department in hopes that the impact on the students' observations and learning experience will be greater.

- "Firehouse Fixins" is our fire department cooking show, televised on cable and a first place winner (fig. 10–34, 10–35).
- "Five Alarms," a cable show about our fire department that has become extremely popular with the public.

Fig. 10–34. Firehouse Fixins host Ronnie Cade with guest cook, Fort Worth Captain Homer Robertson, between takes while filming the cooking show.

Fig. 10–35. It's eating time during this particular show which spotlighted the chuck wagon days and Dutch oven cooking.

- We coordinate the city's annual December Holiday parade (fig. 10–36, 10–37, 10–38).

Fig. 10–36. A great way to showcase the fire department Honor Guard is by leading the Christmas parade.

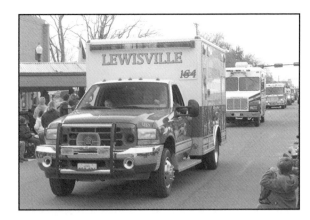

Fig. 10–37. We try to get several fire department units in the parade, which makes for a nice presence.

Fig. 10–38. In Lewisville, the fire department is instrumental in handling the line-up of parade entries and the start and end of the actual parade, a big plus for the fire department with the city powers.

I mentioned earlier that the troops love doing it. Our public service programs have taken on their own life form. The troops come in with all kinds of great ideas. They're the ones out there, working with our citizens, and they get to see just how far their assistance can go. For people who enjoy helping other people (and that's the main reason they became firefighters in the first place) it's a terrific addition to their fire service career.

They just like helping people

And is typical of most firefighters, they do all of the programs I've described here and then some, and then they don't say anything about it. They don't brag. They just have fun helping people. I often get calls from citizens telling me what our troops have done for them. One woman came up to me during an incident at her house and asked to speak with me about one of my firefighters. My first thought was "What did he do?" She began to explain that he was inside folding her clothes. I asked her if he had seen something he wasn't supposed to, but she just smiled and said no. She just thought it was awesome that he would do that for her. A little puzzled, I went in to talk to the firefighter about why he was folding clothes. Sure enough, there was firefighter Chris Kaeppeler, folding clothes. Now, Chris is a big guy with a shaved head and muscles in places where I don't have places. I asked him what had happened. He explained that the washer motor had burned out and they had shut off the power, which in turn shut off the dryer. I said, "Okay, but why are you folding her clothes?" He told me that if he hadn't they would have gotten wrinkled. As goofy as that sounds, he did it because that's what he would have done at home.

Another call I received was from a woman whose neighbor had made a call for help to which we had responded. She explained that her neighbor was 70 years old and mostly bedridden. She had fallen out of bed in the middle of the night and when the firefighters got there she had soiled herself very badly. She didn't need to go to the hospital, but needed to be put back in bed. The woman went on to explain that, before putting her back in bed, the firefighters had changed the linens on the bed, and cleaned and changed her neighbor. By the end of the story the woman was sobbing. She said that she was amazed that our firefighters would treat a total stranger like this and she thought it was awesome. What I told her was that her neighbor was not a stranger to those firefighters. They treated her as if she was their mother, grandmother, or friend. That's how you treat family.

A "sod" story

Recently I received a call from a reporter from *Reader's Digest*. She began to ask me questions about a call during which our firefighters helped a man who was having a heart attack and continued to help with other things after the call was concluded. Once again, I didn't have a clue what she was talking about because the troops rarely come back and brag about helping someone out. They just consider it part of their job.

On this particular call the person having the heart attack had just finished preparing his front lawn to lay new sod. The troops went out and handled the call and went back to quarters. But then their captain, Mark Lee, began to discuss the predicament that the man was in. His lawn would not get done and the sod that had just been delivered would die. I think you know where I'm going with this one. The guys got back on their rigs, went back to the house, stayed in-service, and laid all of the sod. They cleaned up, put the tools away and then commandeered a landscaping company, borrowed their lawnmowers, and mowed the back lawn. Then they quietly went back to quarters and didn't tell a soul. They didn't come back and say "Hey chief, guess what we just did?" They just went out, did it, and went back to the firehouse. *Reader's Digest* called it a "sod" story.

It's gotten to the point that all three shifts try to outdo each other by going the extra mile to help someone with a problem. It's awesome!

Helping someone get back on the right track

My last customer service story is about a car fire. Engine 165 had responded to a car fire and arrived to find one fully involved. Once the fire was out they noticed the owner standing on the side of the road, briefcase in hand and looking a little down, for obvious reasons. Captain Woodward went up to him and asked if there was anything that the fire department could do or someone that they could call for him. The man said no, there was no one he could call and he didn't have family in the area. The captain pressed him and finally the car owner asked if there was a car rental place open so that he could rent a car to get to his next job appointment.

The captain had dispatch call the three rental places nearest their location but found that they were all closed by this time. He didn't stop there. He had dispatch contact the car rental firms at the Dallas-Fort Worth International Airport to see if they could help. Alamo said they could and would love to help them out, so the guys put the owner of the car in their engine, buckled him up, and took him two towns over to the airport to rent a car. When they got there, the car owner got out of the engine, a little puzzled, then asked the captain how much the city of Lewisville was going to charge him. The captain laughed a little and told him that there was no charge. Looking even more perplexed, the owner of the car asked why. The captain told him, because it's what we like to do. They wished him well and off they went.

I know it doesn't sound like much, but if this guy hadn't gotten to his next appointment, he might have lost his job, and if things had gotten bad maybe he wouldn't have been able to pay his bills. Had that happened, it might have taken him a few years of struggle to make things right. Maybe, just maybe, the captain and his crew spun him back onto the right track and kept him from traveling down that road. Think about it. It's not as far-fetched as it sounds.

If you would do it for your family, then do it!

How does the fire department administration support the troops' public service efforts? Simple. We tell them that if it's not illegal, immoral, or dishonest and will help someone, then do it! And if you're wondering whether it's true that "the more stuff you do, the more stuff you get" take a look at our department. The city isn't giving away the store, but our guys are getting raises, they just got 20-year retirement, we've added more people and positions, and for several years now they've gotten pretty much everything that they need and have asked for. It gets better each year. The pay steps for firefighters have just been reduced from eight steps to five, they've approved another ambulance with the staff needed to man it, and to top it all off, we've now got the money for our own bagpipe and drum unit. They've funded the pipes, drums, uniforms, and the lessons! And the best thing of all is that they still get to be firefighters.

I'm still a "truckee" at heart. I still like to go to fires, but I've learned that this customer service stuff really works. It may not work as well everywhere, but it sure does work here. If you don't like the phrase "customer service" then call it something else. I think we'll call it what my dad did about 40 years ago—taking care of family.

Building Relationships

There is so much to be said for building *good* relationships. Unfortunately, there is a lot to be said about building *poor* relationships as well. When it comes to marketing our mission, it doesn't do a whole lot of good for the public to learn that we can't get along with someone or with another agency. Nothing does more to ruin the positive image our hard work has earned us than when a story gets out that one fire department wouldn't respond to assist another or stood by while a building burned, when we're supposed to be getting along and working as a team. The public doesn't understand 20-year-old "feuds" or care that one department doesn't play nice with another. They just assume and rightfully so, that we all get along and work together, as it should be.

> "The public doesn't understand 20-year-old 'feuds' or care that one department doesn't play nice with another."

When you are out there trying to present a good image to the public and get asked questions about why this department didn't respond to assist that department or something along that line, it takes the wind out of your sales pitch sail and ruins a good opportunity. I'm still amazed today when I see a fire in our area and hear that certain departments didn't respond or were never requested even though they are located in the next town over. Does it really matter what color your helmet or engine is or who didn't get along with whom years ago? When it comes to that kind of thinking, forget about the customer; think about the safety of your personnel. Don't let an ego run amuck jeopardize the safety of your firefighters. Remember one of our favorite sayings, "Egos eat brains." Ego and the kind of problems it creates make it harder for you to get all the things you want.

That whole PD vs. FD thing again

The same goes for how well you get along with your police department and any other city departments. I remember having a rough start here when it came to our relationship with the public works department. Because of some issues that occurred before I got here, we wasted a couple of years before getting everything back on track. We have a great relationship with the public works department today.

We all need each other. Our training for large-scale disasters has certainly demonstrated the need to get along and work well with each other. There are some big city fire departments who now realize they can't do it on their own and really do need to rely on their neighbors. And that's not a bad thing!

Working with City Hall

It's also important to have a good relationship with the decision makers at City Hall. Whether it's the city manager, mayor, or city council, how well do you get along with the people who hold the purse strings? What kind of a relationship do you have? If you've ever wondered why you never get anything in the budget process or get beat down on an idea no matter how good it seems to be, take an honest look at your relationship with your bosses and you'll probably have your answer. I know you may feel that some relationships just can't be built (probably because the person you're dealing with has a head made of stone!), but it's worth a try. You might be surprised at how it turns out. Be adult and professional. Don't allow yourself to be brought down to their level because those on the outside will notice.

Just as important is building and strengthening relationships with your neighboring fire departments. There's no good excuse for not working with each other, or at the very least, trying to work together.

Work with who has the "pull"

Lastly, remember to work hard at building relationships with service organizations, home owners associations, schools, PTAs, and anyone else who will listen. Don't forget seniors groups. They've paved the way for a lot of things, usually know everybody, and often have a lot of pull when it comes to getting things done. It doesn't take a lot of work or time to put a presentation together. If you're really short on time or they can only give you a short amount of time, try the following idea, which I got from Bruno. Take some blank index cards and when you're being introduced and doing your lead-in, have someone from the group, maybe the person who invited you, hand them out to the audience. Ask the people in the audience to write down the one thing that they want to see from their fire department. What's the most important thing they want? Have the index cards collected and have someone read each one to you. Most of them will say things like, "protect me and my family," and "respond quickly and put out my fire." You'll also see various things like "make my insurance payments go down" and so on. When you respond to each comment you'll find that your audience has written your program for you and, at the same time, gotten answers to their questions. Nine times out of ten you'll run out of time.

When it comes to marketing our departments and what we do, the roadblocks we run up against were often set up by some of our own people. It just takes some extra work to push them out of the way and work toward ways to get things done. Fight for your community's support, the support of those at city hall, and of the other city departments and do so by leading by example. Be the first to "lay down your arms" and offer a way to get along. Be innovative when it comes to selling an idea. Reasons based on safety or liability will at times lose their power and you will need to find another avenue.

Seize the moment

Go out and brag about what we do and I mean *all* that we do. There's nothing wrong with going out and telling people that you care or explaining how you take care of folks. Seize the moment, as Matt and David did in Atlanta. They were pretty smart thinkers, if you ask me!

Be everywhere. The cops are. Get involved and try to be included in every event, every function. And, Chief, let the guys go shopping for their meals. It's not a waste of the taxpayer's money and it gets us out with the public 365 more times a year.

Take the stand that we're not going to go away and that we do what no else can do as well. Fight to keep what we have. Don't let it slip away by sitting back and failing to get involved. So many have fought and bled for us over the years to get us where we are today. Don't let what they did be wasted. *Go out and brag about the fire service*!

> "Go out and brag about what we do and I mean *all* that we do."

11

MAKING IT ALL HAPPEN AND TAKING CARE OF NUMBER ONE

Throughout the past ten chapters, while we discussed pride and ownership, we have again and again discussed just how great this profession of ours is. We started by taking a look at our *mission*, at why we do what we do, and at what the fire service is all about. Then we looked at the kind of person it takes to carry out our mission, to be a *firefighter*, and the devotion and commitment it takes to be good at what we do. We said that if you don't like it; if you're not willing to take care of each other and defend our family, then get out! We realized that to ensure that our firefighters could do what they needed to do and grow within their careers, they need someone to guide them, mentor them, and build them a path for success. We called that someone the *company officer*.

All good teams need good coaches. Our team needs people who can work with the company officers to help and mentor firefighters in the best way possible. At that point we talked about our *battalion chiefs* and what they do for all of us. Knowing who really makes it happen in our business—those firefighters, company officers, and battalion chiefs—we also know that, in order for them to succeed as a team, to really make a difference, they need the support of someone who understands and appreciates them; *the Chief!* A true chief cares more about the troops, their staff, and the overall mission than they do about themselves. Finding that kind of person can be difficult. But the same is true of the firefighters, company officers, and battalion chiefs we talked about.

We've discussed those that work against us and our mission, but in doing so, realized that they are few and far between. The majority of the people in our family are good, hard working, and dependable and want nothing more than to make a difference in other people's lives. But boy does it get really hard to keep a lid on this whole thing when you have a mutt for a chief who just doesn't care. That whole "It's hard to soar with eagles when I'm surrounded by turkeys" saying, at times needs to say as a chief, before I look for eagles, I need to look in the mirror and see what kind of feathers I'm wearing. Now, with that kind of accountability and understanding at the top you quickly end up saying, "Man, look at all that we're doing and getting done." And that kind of environment gets you quickly to where you're saying there is no telling what we're going to get done and that the sky's the limit.

The nature of our work and the ways in which we interact with one another make the people in the fire service a very close-knit family. Our fire service family is like no other, though it's a goal many strive for, especially in the private sector. And we have adjusted to having two families, *Our Two Families,* one at work and one waiting at home by making the family at home the first family and by becoming protective of both. Then we took a hard look at some of the areas that have given us problems, hurt us, or ,worse yet, killed us. We learned that while others tell us we shouldn't be *sweating the small stuff* it's the small things that we overlook that always come back to haunt us.

Next we talked about the challenges of preparing for, taking, and passing the promotional process and then about getting promoted, or *changing shirts,* and making the very difficult adjustment from buddy to boss, learning how to supervise, leading those looking up to you and making decisions and choices on their behalf because it's the right thing to do and not because it's the most popular choice. Later we took an honest look at what *9/11 did to us and for us,* and emphasized that "NEVER FORGETTING MEANS NEVER FORGETTING!" We talked about the *ceremonies* that do us proud, are well-deserved, and are needed to preserve our heritage. And finally we talked again about *marketing our mission*, our departments, ourselves, and what we're all about. Especially if were going to survive in today's economy.

Are You Being Honest with Yourself?

But when we take a good look at all of these things, you get to a point that in order to truly make it all happen, really happen, and take care of number one, you have to care, work hard, set your priorities, and be brutally honest with yourself. Are you living up to your expectations and of those around you? Because its not just a matter of what the boss expects of me but what those next to me and below me expect. Only then will you know that you are taking care of number one.

Sounds kind of selfish doesn't it? Firefighters thinking of themselves first. It doesn't seem to fit, or does it? In order to commit to a life of selflessness and be good at it, you have to take care of yourself first. Health wise, mind wise, at home and at work, and continue to do so throughout your life. Someone who runs him or herself into the ground or doesn't take care of themselves and forgets about their priorities will struggle in the long run. At the firehouse, how good are you? How much do you know? How much are you willing to learn and what energy level are you working at, low, medium, or high? When we dip to the low side we suffer and bad things tend to happen. Remember how much attitude played into it.

"Are you living up to your expectations and of those around you?"

Embracing Success

You have to set goals and prioritize them. Identify the steps or objectives that will get you where you want to go. Celebrate when you succeed. Some people have never been shown how to realize they are successful and how to embrace that success. On several occasions a firefighter or company officer has walked into my office, sat down, and started to explain that they feel like they're in a rut, like they're not contributing. Often they are forgetting to look at how much they have achieved toward the goals they set. Hopefully, if their goals were attainable, it's just a matter of realizing where they have been successful.

We get so wrapped up in our jobs and are in such a hurry to get this or that done and move on to the next project that sometimes we get to where we wanted to go without even realizing it! If that's the case with you, give yourself a break and a chance to catch up and enjoy. Embrace success!

Stop and Smell the Roses

Before you can embrace success, you have to be able to recognize it. Once you have, and have taken the time to enjoy it, catch your breath, reexamine your current list of goals and objectives, reprioritize them, and move ahead. But don't forget to enjoy the fruits of your labor. Stop to smell the roses just for a minute. Remember what success feels like and that it's okay to feel good about it. Otherwise you'll get frustrated, feel empty, unappreciated, and as if you're stuck in a rut. Eventually you'll burn out. It's similar to what happened to a lot of folks when they went to direct deposits with their paychecks. When the money didn't pass through their hands, when they didn't see a return from their hard work, they felt like they hadn't accomplished anything. They say it leaves some people with a void, an empty spot, almost like they didn't finish something.

> *"Amateurs built the Ark — Professionals built the titanic."*

Learn from Your Successes the Same as You Do Your Failures

I'll tell you one last thing about embracing success. You have to learn from your successes just as you do your failures. We sometimes forget that and focus our attentions on our failures. But failing to learn from success can lead to mediocrity down the road. Unless we examine what made us successful in the past, we don't know where to set the bar for improvements or we may not be able to apply the same strategies and methods to other goals. We need to look at both failures and successes to keep moving forward in the right direction.

Stay fresh. Hang around those with good attitudes. Continue to make your mark in the fire service, and on a regular basis say to yourself, "Wow, what this place could be if…" and see what you can do to get it there.

> "Wow, what this place could be if…"

12
HAVE YOU FORGOTTEN

As we wind down to the end of our discussion of pride and ownership, I can't think of a better time to talk about the idea of never forgetting. We've mentioned it several times in past chapters but it definitely seems to be worth mentioning again. Several months ago, I was invited to give a keynote speech at a fire department conference and was graced with the presence of one of that city's aldermen. The host chief had invited him to speak at the opening ceremonies and had brought us all together the morning of the conference for breakfast. We sat and visited about different issues and somehow got around to the topic of funding within the fire service, and in many areas of the country, the lack of it. We talked about how law enforcement seems to have done well and that we just wanted our share. It doesn't have to be that big, just something.

Using 9/11 as a Crutch

Well, we must have pushed the right button because the alderman started going off about the fire service, saying that all we want to do is spend money, hire more firefighters, and build our own little kingdoms. Yes, he went on, what happened on September 11, 2001 was a terrible tragedy and a horrible loss, but people shouldn't lose focus on reality and what is reasonable. It's not right to use that event to emphasize a specific need or want, he added. Those things just don't happen that often and we shouldn't use 9/11 as a crutch.

Now, as you can imagine, I was starting to rise off of my seat, just as I'm sure you would have. I could feel my neck getting hot and my eyes starting to burn. But I kept reminding myself to keep my cool and not say anything that I would regret later or anything that would embarrass the host chief. I could tell that he was feeling bad enough already about what his alderman was saying. So I decided to bite my tongue and did until it bled, all along thinking about and planning how I was going to play this one out and make a point with this guy. I knew that arguing with him at breakfast was going to go nowhere. Arguing with someone like him is about as useful as shouting at the rain. He just wouldn't get it.

So I waited until it was time for my keynote speech. I decided that the speech I was supposed to give was going to have to wait and I was going to go with something a little different. I asked the chief who was introducing me to change the title of my speech from *Pride and Ownership; the Love for the Job*, to *Have You Forgotten?* And, as it happened, I got my chance to speak right after the alderman.

The next 35 minutes were very interesting. I began talking about what happened on September 11, 2001, and about the loss we all felt when we lost so many firefighters, not to mention the pain our brothers in New York felt. I talked about losing that much talent and so many special people. We looked back at all of the ceremonies and fundraisers and at the effort put forth by not just the fire service, but by our entire country. We also talked about *some* of the pictures and handshakes with all of the politicians back then. Notice I said some, not all! I recognize that there have been many good politicians who have fought and continue to fight for us and for what we need. They truly do care about us.

But then there are the politicians like the good alderman. After all that he'd said that morning, he still stood up there and talked about his great relationship with the fire service. Great relationship? What in the world was he talking about? I was there, and I heard him at breakfast. Wasn't this the same person that said all we want to do is build our little kingdoms and spend money? Didn't he refer to the fire service as a dark hole where all the money goes? Wasn't this the same guy that said we should stop using 9/11 as a crutch to get something? Wasn't he one of them standing next to us in the pictures and putting his arm around firefighters? He sure was! I felt like we were being used as a public relations stunt to get votes. I was astounded that he could have forgotten—forgotten everything. Then it hit me. I'd been thinking that he said those things about the fire service and about 9/11 being a crutch because he had forgotten. But he hadn't forgotten about the courage and selflessness shown by our family that day. He'd simply never believed in it in the first place.

I continued with my speech mentioning that some have found it easy to forget, but stayed calm and professional all the way. Well, certainly professional. I got a little loud once in a while. But I didn't stop there. I didn't stop with 9/11.

Never Forgetting Means Never Forgetting

I talked about the fire chiefs, and, yes, even some of the firefighters who had forgotten and I couldn't help thinking that *Never Forgetting Means Never Forgetting."* There were chiefs and city managers who fought us on staffing issues, training needs, and over the need for safe equipment and apparatus. But after 9/11 they disappeared, at least for a while. They're back now, and they're fighting us on the same issues again. And they will continue to do so until the next big crisis. Then they'll disappear again. The politicians will be back, though, for pictures and handshakes. But I won't be standing next to them next time. I will be next to the people who support and help the fire service. I'll be there for them as they have been for us.

The fire service needs to maintain constant vigilance to assure that we receive a fair share of the federal and state funding that is being allocated for things like Homeland Security, rather than seeing it tilted to one side. The fire service needs to make their voices heard by letting our elected and appointed leaders know that response is just as important as security.

Forgetting Just One Is One Too Many

Are our own people, our firefighters, officers, and chiefs forgetting not just those we lost on 9/11, but those we lost before and have lost since? Never forgetting means never forgetting any of them. Forgetting even one is one too many. Have we forgotten what they meant to us? What they did for us? How they affected our lives? We've lost so many special people. I know of a department that lost three firefighters in two different fires and yet you'd be hard-pressed to find a picture of them anywhere. They found it easy to forget.

So much talent has been taken away from us. Are we willing to forget what it took to get us to where we are today? What it took to get us to where we are with equipment, training, and safety? Pretty much all of it is due to someone else's hard work, sweat, blood, and, in some cases, lives. Why does it seem so easy to forget? So many special people worked hard to get us here. How can we forget them? Maybe if we shared what went wrong at some of the fires that resulted in a firefighter fatality, we could really start to make a difference in the numbers and keep from losing more of our own. Maybe by doing that we can prevent history from repeating itself. But in order to do so, we have to be honest and up-front about what happened. Maybe, just maybe, we can make the same kind of difference that Dave McGrail did in Denver, Colorado. And, no, we shouldn't let it go or move on. We owe it to those we have lost and to their families to learn from their deaths and see that it doesn't happen again.

Back to That Leadership Thing Again

A lot of it reflects back to our leadership—or lack of it—and their inability or unwillingness to make us remember. To be fair, there are some out there fighting to keep our memories fresh and keep people remembering the sacrifices made by others. They are helping to explain where it all started and working to define our heritage. They are still conducting ceremonies and showing one another honor and respect. They remember to hold a ceremony on September 11 each year (fig. 12–1, 12–2, 12–3, 12–4, 12–5). But there are still members of our family out there who don't have the foggiest idea where the fire service started and what we're all about. They don't have a clue about our history, our traditions. If they're not willing to learn about it and share it, it's going to erode and wash away. Eventually we'll lose it forever.

Fig. 12–1. Members of the fire department line up for the 9/11 memorial.

Fig. 12–2. The dedication of the memorial was attended by many from the public.

Fig. 12–3. Patriot Day now allows us to fly the flag at half staff on 9/11 to honor those we lost that day.

Fig. 12–4. Bell ringing has been part of memorial services for decades and adds another touch of honor to the presentation.

Fig. 12–5. The memorial is open to the public year round and is a place for firefighters to sit and reflect.

But then again, maybe it's more than just the job of our leadership to keep the fire service heritage alive. Maybe we need more of our officers and firefighters to remind others of where it all came from. If you're a firefighter and don't know, ask someone who does and then go out and spread the word.

Maybe fewer people will forget then. Never forgetting means never forgetting. Maybe we need more classes and articles on fire service history. We might even come up with a method for not forgetting. Maybe it's time some people checked their egos, stopped worrying about how much more someone else knows, and started worrying more about being good at what they're already doing. I'm not saying that you should stop trying to better yourself or improve in an area, quite the opposite. I'm just saying we need to be a little nicer to each other and never forget what it means to be a brother or what it took to get us all here. Never forgetting means never forgetting.

Don't Confuse a Tribute with Honoring Someone

We also need to remember the difference between paying tribute to someone and honoring someone. When we say goodbye to a fallen comrade or to one who has passed, it's time to pay them a *tribute,* to acknowledge what they've done for us, and to thank them for their friendship. The time to *honor* them was before they left us. We should be honoring each other while we're all still here and still have a chance to really let people know what they mean to us and how much we appreciate them. Let's not wait until after they're gone. The time for honor is right now. Do that and you can truly say you're a brother. Maybe we need a few more ceremonies to honor and celebrate someone's accomplishments, a few more ways to remember. Maybe that will help. I really don't want to wait until another life to tell and show people just how much I appreciate what they have done for me or for us.

I messed up that opportunity with my little brother and I don't want to do the same with a friend. I don't want to just look up to the sky and say thanks. I've missed too many opportunities. I'm not going to miss any more. Never forgetting means never forgetting. Our fire service family is way too special to forget about anyone or anything that got us to where we are today. Do you remember your retirees? Their names? What they did for you and how they made it a little better for all of us?

The Fire Service Is the Greatest Profession in the World

I don't know. Maybe it's a lot to ask for. Maybe it isn't. I'd like to think it's worth trying a little harder. The fire service is the greatest profession in the world. I'm never going to forget about those that we've lost. And if that means that the alderman who had a lot to say at breakfast will think I'm using them as a "crutch" to make our fire service better and safer for the next guy, so be it. I know they'd say go for it. See, they can really still work for us and help us even if they're gone. If we're willing to never forget. When I was finished with my speech, the alderman came up to me and said that it had been great and had really hit home. I thought, yeah, for about five minutes with you.

In closing, it's time for one more shot. If you don't love this job, if you're not willing to stand up for and protect a brother or sister firefighter, if you're not willing to protect our fire service family and its image, it's simple: get out. Leave, go away. Because what we really need are more people who feel pride and ownership and love for the job. Man, I love being a firefighter!

Appendixes

Note from the Author

The following sections are a compilation of frequently requested documents that the Lewisville Fire Department has created and uses on an everyday basis.

I have divided the documents into sections based on their area of use: Mentoring, Departmental Communications, Ceremonies, After the Fire customer support information, and Miscellaneous Forms.

I hope that these forms can be used to help organize the paperwork side of the job. Please feel free to use these templates for your own fire department and modify them to fit your needs.

MENTORING

Mentor Questionnaire

GOAL: In an effort to capture as much information as possible from our senior members, *our mentors*, before they leave us, the Lewisville Fire Department has developed the "Mentor Questionnaire." You are being asked to take a few minutes and share your KNOWLEDGE, EXPERIENCES, and TRICKS OF THE TRADE, with our new Firefighters. This information will allow us to pass along your experience and our Department's heritage to our newest and future members.

INSTRUCTIONS: KEEP IT SIMPLE! You don't have to answer them all but it would be greatly appreciated. They can be typed *or* hand written. Don't worry about misspelling or grammar. Those haven't saved any firefighter's lives, but maybe with this you can!

NAME: _____ RANK: _____

QUESTION 1 – (Apparatus, Tools & Equipment)

What can you tell us about our apparatus? Are there any with special needs or "quirks?" Any safety concerns that you would like to pass along? This can also apply to equipment on our rigs such as extrication equipment, generators, saws, tools, etc. Is there a rig you like best and why? Is there a rig you like least and why?

QUESTION 2 – (Calls, Incidents, Experiences or Problem Buildings)

Are there any Calls, Incidents, or Experiences that you would like to share? Whether for historical purposes or more importantly, a call or incident that provided you with a "Lesson Learned" type situation that may help a new Lewisville Firefighter. Are there any buildings that you are concerned about? The type of building that when you think about it you find yourself saying "I hope we never have a fire in that place."

QUESTION 3 – (Lake Lewisville or Dive Operations)

Is there anything specific about Lake Lewisville or our Dive Team Operations that you would want to share with a new Lewisville Firefighter? Any hazards or safety concerns that would help keep one of our members out of trouble?

QUESTION 4 – (Department History)

Can you provide us with any information regarding our Department's History? Anything you would like to share with future generations of Lewisville Firefighters? History regarding our firehouses, apparatus, personnel, incidents, etc?

QUESTION 5 – (Your Wisdom and Thoughts)

Is there anything that you would say to a new Lewisville Firefighter that would help him or her stay healthy and safe throughout their career?

Mentor Book
Table of Contents

Mentor Questionnaire...Question 1
Apparatus, Tools, Equipment

Mentor Questionnaire...Question 2
**Calls, Incidents, Experiences,
Problem Buildings**

Mentor Questionnaire...Question 3
Lake Lewisville, Dive Operations

Mentor Questionnaire...Question 4
Department History

Mentor Questionnaire...Question 5
Your Wisdom and Thoughts

Step-up Process
Firefighter to Driver Engineer

Step-up Process
Driver Engineer to Captain

Step-up Process
Captain to Battalion Chief

New Members

Miscellaneous

Step-up Process

Firefighter to Driver Engineer

- Recommendation by Captain
- Written examination
- Practical exercise

Driver Engineer to Captain

- Written exam
- In-Basket Exercise
  - Accident investigation and report
  - Injury investigation and report
  - Employee issue
  - City / department policy application
  - Written communication
- Tactical exercise
  - One alarm residential

  - First-on
  - First 10 minutes
    - Size-up
    - Initiate IMS
    - Develop an action plan
    - Identify problems
    - Evaluate resources
    - Deploy resources
    - Safety considerations
    - Communications
    - Transfer of command
- Field mentoring
- Captain's recommendation
- Battalion Chief's endorsement
- Training Chief endorsement
- Assistant Chief / Operations approval

(Step-up Process continued)

Captain to Battalion Chief

- In-Basket Exercise
  - Accident investigation and report
  - Injury investigation and report
  - Staffing issue
  - City / department policy application
  - Written communication

- Tactical exercise
  - Two alarm multi-family residential/commercial strip
    - Size-up
    - Assume command
    - Expand IMS
    - Develop a strategy
    - Address incident priorities
    - Evaluate problems
    - Evaluate resources
    - Safety considerations
    - Communications
    - Terminate command

- Shadowing

- Field mentoring

- Battalion Chief's recommendation

- Training Chief endorsement

- Assistant Chief / Operations approval

Step-up Process Form

Employee's Name_____

Step-up Process: **Driver Engineer**   **Captain**   **Battalion Chief**

_____

_____   Letter of recommendation from Captain

_____   Letter of recommendation from Battalion Chief

_____   Written exam   PASS   /   FAIL

_____   Driving Practical   PASS   /   FAIL

_____   Pumping Practical PASS   /   FAIL

_____   In-Basket   PASS   /   FAIL

_____   Tactical Exercise   PASS   /   FAIL

_____   Assistant Chief of Operation's Approval

_____
Signature Assistant Chief – Operations

_____
Signature Training Chief   Date

Officers Academy Part I

Officer I	Core Values
	Roll Calls
	Shift Meetings
	Seat Assignments
	Mentoring/Preparing to Promote
	Shift-to-Shift Communications
Chief Tittle (45 min.)	Notifications
	Area Assignments
	Station / Safety Inspections
	Hydrant Program
Chief Nolen (20 min.)	Uniforms
Chief Thompson (20 min.)	Daily Training Records
	Injury / Accident Reporting
	Mentor Programs
	New Hire Checklist

Officers Academy Part II

Course	Dates	A.M. Session	P.M. Session	Facilitator	Topic
Officer II	**February 24,25,26**	8:30–9:00	1:00–1:30	**Chief Nolen**	Budget Process
(All Chiefs, Captains, Driver Engineers)					ISO Overview
Copies of Handouts Needed (52)		9:00–10:00	1:30–2:30	**Chief Tittle**	Awards / Recognition Performance Evaluations
		10:15–11:15	2:45–3:45	**Chief Lasky**	Disciplinary Issues Career Development
		11:15–11:45	3:45–4:15	**Chief Thompson**	Promotional Process New Hire Orientation Part 2 Pre Incident Planning

Officers Academy Part III

Course	Dates	A.M. Session	P.M. Session	Facilitator	Topic
Officer III (All Chiefs, Captains, Driver Engineers) **Copies of Handouts Needed (57)**	March 24,25,26	8:30–11:30	1:00–4:00	**Battalion Chief**	Engine Operations
					Truck Operations
					Brush Fire Operations
					Water Supply
				Battalion Chief	Incident Management
				Chief Nolen	Radios Communications
				Chief Thompson	Simulator

Officers Academy Part IV

Course	Dates	A.M. Session	P.M. Session	Facilitator	Topic
Officer IV (All Chiefs, Captains, Driver Engineers) **Copies of Handouts Needed (57)**	April 21,22,23	8:30–11:30	1:00–4:00	**Chief Thompson** Station 2	Special Team Responses Dive
				Irving FD	Haz Mat
				Chief Thompson	
				Coppell FD	Trench
				"	Confined Space High Angle
				Carrollton	Swift Water
				Chief Nolen	Mutual Aid

Officers Academy Part V

Course	Dates	A.M. Session	P.M. Session	Facilitator	Topic
Officer V	July 28,29,30	8:30–11:30	1:00–4:00	Chief Lasky	Customer Service
(All Chiefs, Captains, Driver Engineers)					
				Chief Nolen	Policy Review LFD City
Copies of Handouts Needed (57)					

Officers Academy Part VI

Course	Dates	A.M. Session	P.M. Session	Facilitator	Topic
Officer VI	August 25,26,27	8:30–11:30	1:00–4:00	Chief Shade	Apparatus
(All Chiefs, Captains, Driver Engineers)					New Equipment
Copies of Handouts Needed (57)				Chief Wilkins	Investigations
				Chief Kohn	Emergency Management EOC Sirens

Officers Academy Part VII

Course	Dates	A.M. Session	P.M. Session	Facilitator	Topic
Officer VII	Sept. 29,30	8:30–11:30	1:00–4:00	Chief Lasky	Managing the MAYDAY
(All Chiefs, Captains, Driver Engineers)	Oct. 1				Rapid Intervention
Copies of Handouts Needed (57)					

Notice of Intention to Enter Promotional Selection Process

Date of Filing: _____, 20_____

Time: _____

I hereby give notice of my intention to enter and compete in the promotional selection process for the position of _____.

On the date scheduled for the examination, I will have completed _____ years and _____ months continuous service in the Lewisville Fire Department at the rank of _____.

_____
Applicant's Name (Print)

_____
Applicant's Signature

_____
Human Resources Staff

Promotional Exam for Engineer

Date: January 20, 2005

To: All Personnel

From: Richard A. Lasky, Fire Chief

Re: Reading List for the Promotional Examination of Driver/Engineer

Please find listed below the reference material that will be used for the written examination.

The Training Division has placed a copy of each reference being used for the written examination in Stations 2 through 6 and two copies in Station 1. These are to be utilized by on-duty personnel or those that choose to study in the Station on off-duty time. At *no* time shall any of these references or copies already in Stations leave their assigned Station or be stored in a personal locker. Please see that when you are finished with a particular copy, that it is returned to the Station library. This will help ensure that all personnel have access to the reference materials.

Pumping Apparatus Driver/Operator Handbook, 1st Edition, IFSTA, 1999

Aerial Apparatus Driver/Operator Handbook, 1st Edition, IFSTA, 2000

Fire Department Company Officer, 3rd Edition, IFSTA, 1998

(STUDY Chapters 1, 3, 4, 5, 6, 7, 13, 14, 17, 20, 21, and 22 ONLY)

Fire Command, Alan V. Brunacini, 2nd Edition, Heritage Publishers, 2002

Safety and Survival on the Fireground, Vincent Dunn, 1st Edition, Fire Engineering, 1992

Promotional Exam for Captain

Date: January 20, 2005

To: All Personnel

From: Richard A. Lasky, Fire Chief

Re: Reading List for the Promotional Examination
                 of Captain

Please find listed below, the reference material that will be used for the written examination, along with the information that will help you prepare for the tactical and in-basket exercises.

The Training Division has placed a copy of each reference being used for the written examination in Stations 2 through 6 and two copies in Station 1. These are to be utilized by on-duty personnel or those that choose to study in the Station on off-duty time. At *no* time shall any of these references or copies already in Stations, leave their assigned Station or be stored in a personal locker. Please see that when you are finished with a particular copy, that it is returned to the Station library. This will help ensure that all personnel have access to the reference materials.

Building Construction for the Fire Service, Francis L. Brannigan, 3rd Edition, NFPA, 1992

(STUDY Chapters 2, 3, 4, 9, and 13 ONLY)

Fire Officer's Handbook of Tactics, John Norman, 2nd Edition, Fire Engineering, 1998

Fire Command, Alan V. Brunacini, 2nd Edition, Heritage Publishers, 2002

Safety and Survival on the Fireground, Vincent Dunn, 1st Edition, Fire Engineering, 1992

Fire Department Company Officer, IFSTA, 3rd Edition, 1998

(STUDY Chapters 1, 3, 4, 5, 6, 7, 13, 14, 17, 20, 21, and 22 ONLY)

Essentials of Fire Department Customer Service, Alan V. Brunacini, 1st Edition, FPP, 1996

Supporting reference material for the tactical and in-basket exercises are as follows:

City Fire Department, Operating Guidelines, Sections 3, 4, & 5.
City Policy and Procedure Manual, Section 2.

Promotional Exam for Battalion Chief

Date: January 20, 2005

To: All Personnel

From: Richard A. Lasky, Fire Chief

Re: Reading List for the Promotional Examination of Battalion Chief

Please find listed below, the reference material that will be used for the written examination, along with the information that will help you prepare for the tactical and in-basket exercises.

The Training Division has placed a copy of each reference being used for the written examination in Stations 2 through 6 and two copies in Station 1. These are to be utilized by on-duty personnel or those that choose to study in the Station on off-duty time. At *no* time shall any of these references or copies already in Stations, leave their assigned Station or be stored in a personal locker. Please see that when you are finished with a particular copy, that it is returned to the Station library. This will help ensure that all personnel have access to the reference materials.

Building Construction for the Fire Service, Francis L. Brannigan, 3rd Edition, NFPA, 1992

(STUDY Chapters 2, 3, 4, 7, 8, 9, 12, and 13 ONLY)

Fire Officer's Handbook of Tactics, John Norman, 2nd Edition, Fire Engineering, 1998

Fire Command, Alan V. Brunacini, 2nd Edition, Heritage Publishers, 2002

Safety and Survival on the Fireground, Vincent Dunn, 1st Edition, Fire Engineering, 1992

Essentials of Fire Department Customer Service, Alan V. Brunacini, 1st Edition, FPP, 1996

Supporting reference material for the tactical and in-basket exercises are as follows:

City Fire Department, Operating Guidelines, Sections 3, 4, & 5.
City Policy and Procedure Manual, Section 2.

Firefighter Core Competencies

Employee Name		Date	3/13/06
Social Security #		Job Title	Firefighter
Time Period Being Evaluated		to	
Department/Division	Fire/Operations		

Emergency Medical Scene

- Does the employee effectively apply EMS knowledge and skills during emergency situations, staying calm and using good judgement during stressful medical situations?

Competency: Meets Standards

Comments:

Commitment to Public Service

- The employee consistently deals with the public in a prompt and considerate manner, seeking out the appropriate information or individual to meet the citizens' needs.
- The employee positively communicates regarding the organization's programs, goals, policies, etc.

Competency: Meets Standards

Comments:

Street Knowledge

- Employee participates in quarterly street tests and maintains a 80% average.

Competency: Meets Standards

Comments:

Behavior and Teamwork

- Employee is courteous to the public, administration, and co-workers.
- Employee works openly and positively to accomplish city and department goals.
- Employee offers tactful and constructive suggestions.
- When dealing with administration or co-workers, the employee establishes, promotes, and maintains a positive working relationship.

(Firefighter Competencies continued)

- Employee demonstrates knowledge of Chain of Command, Operating Guidelines/Rules & Regulations.

Competency: Meets Standards

Comments:

Reports (Fire, EMS, Daily Reports)

- The employee completes computer reports assigned to him by the end of shift. Not filling out Texas Fire or EMS Report prior to the end of shift without supervisor's permission twice in one year constitutes an oral reprimand.

Competency: Meets Standards

Comments:

Fire Scene

- The employee shows good judgement at the fire scene while following all safety guidelines.
- The employee shows willingness to participate in all fireground activities assigned, follows directives given, and stays focused on task.

Competency: Meets Standards

Comments:

Knowledge of Fire and EMS Equipment

- The employee demonstrates working knowledge and location of all fire and EMS equipment at his/her assigned station.

Competency: Meets Standards

Comments:

Equipment Check and Daily Assignments

- The employee thoroughly checks apparatus, notes and reports any items missing or broken. Completely marks checksheet, makes minor repairs such as changing light bulbs, etc., and reports any needed repairs.
- The employee takes an active role in completing the daily duties and completes all tasks assigned by the station officer.

Competency: Meets Standards

(Firefighter Competencies continued)

Comments:

Experience Raise (*Value = 50% including written and skill test*)
- Employee satisfactorily completed and passed said experience test.

Competency: Meets Standards

Comments:

Signature of Employee	Date
Signature of Supervisor	Date
Signature of Assistant Chief	Date
Signature of Fire Chief	Date

Firefighter Compensation Summary

Employee Name		Date	01/26/2005
Social Security No.		Job Title	
Current Step		Date of Hire/Promotion	
Department/Division			

If the employee is at the top step and is not eligible for an increase, check only those that apply in Section I. If an employee is eligible for a step raise, complete both Sections I and II. Forward the original of both this form and the Core Competencies Evaluation to Human Resources.

I. Overall assessment of employee's performance: *(check all that apply)*

	a)	Met all core competencies for past year.
	b)	Did not meet standards on two competencies or received one written reprimand related to one ore competency.
	c)	Passed written test.
	d)	Did not pass written test.
	e)	Passed skills test.
	f)	Did not pass skills test.

II. Step increase determination

	Eligible to move to step		*must have checked a), c), and e)*
	Not eligible at this time		
	Retest <u>date</u>		*if checked d) or f) above*
	End of performance improvement plan <u>date</u>		*three months after date of evaluation*
	Raise withheld until satisfactory completion of PIP or test		

Signature of Employee	Date
Signature of Supervisor	Date
Signature of Assistant Chief	Date
Signature of Fire Chief	Date

Driver Engineer Core Competencies

Employee Name		Date	01/26/2005
Social Security #		Job Title	Driver Engineer
Time Period Being Evaluated		to	
Department/ Division			

Emergency Medical Scene

- Does the employee effectively apply EMS knowledge and skills during emergency situations, staying calm and using good judgement during stressful medical situations?

Competency: Meets Standards

Comments:

Commitment to Public Service

- The employee consistently deals with the public in a prompt and considerate manner, seeking out the appropriate information or individual to meet the citizens' needs.
- The employee positively communicates regarding the organization's programs, goals, policies, etc.

Competency: Meets Standards

Comments:

Street Knowledge

- Employee participates in quarterly street tests and maintains a 70% average.

Competency: Meets Standards

Comments:

Behavior and Teamwork

- Employee is courteous to the public, administration, and co-workers.
- Employee works openly and positively to accomplish city and department goals.
- Employee offers tactful and constructive suggestions.

(Driver Engineer Core Competencies continued)

- When dealing with administration or co-workers, the employee establishes, promotes, and maintains a positive working relationship.
- Employee demonstrates knowledge of chain of command, operating guidelines/rules & regulations.

Competency: Meets Standards

Comments:

Reports (Fire, EMS, Daily Reports)

- The employee completes computer reports assigned to him by the end of shift. Not filling out Texas Fire or EMS Report prior to the end of shift without supervisor's permission twice in one year constitutes an oral reprimand.

Competency: Meets Standards

Comments:

Fire Scene

- The employee shows good judgement at the fire scene while following all safety guidelines.
- The employee shows willingness to participate in all fireground activities assigned, follows directives given, and stays focused on task.

Competency: Meets Standards

Comments:

Knowledge of Fire and EMS Equipment

- The employee demonstrates working knowledge and location of all fire and EMS equipment at his/her assigned station.

Competency: Meets Standards

Comments:

Equipment Check and Daily Assignments

- The employee thoroughly checks apparatus, notes and reports any items missing or broken. Completely marks checksheet, troubleshoots, and makes minor repairs. Reports any repairs that cannot be readily made.
- The employee oversees and participates in daily duties.
- The employee completes all special projects assigned by station officer.

(Driver Engineer Core Competencies continued)

Competency: Meets Standards

Comments:

Resource Management

- The employee organizes work to use time efficiently and effectively, relating and arranging workload and resources (equipment, facilities, and personnel) and delegates responsibilities to meet desired results.
- In captain's absence, the employee assists with area of responsibility.

Competency: Meets Standards

Comments:

Experience Raise *(Value = 50% including written and skill test)*

- Employee satisfactorily completed and passed said experience test.

Competency: Meets Standards

Comments:

Signature of Employee	Date
Signature of Supervisor	Date
Signature of Assistant Chief	Date
Signature of Fire Chief	Date

Captain Core Competencies

Employee Name		Date	02/8/2006
Social Security #		Job Title	Captain
Time Period Being Evaluated		to	
Department/Division			

Employee Development

- Employee assists in the development of his/her personnel through on-the-job training.

Competency: Meets Standards

Comments:

Area of Responsibility

- Employee effectively manages area of responsibility. This includes submitting budget in a timely manner, and effectively managing account balances throughout the year.
- Employee is innovative in maximizing the use of funds in his/her budget.

Competency: Meets Standards

Comments:

Resource Management

- Employee organizes work to use time efficiently and effectively.
- Employee completes work assignments on time.
- As a supervisor, employee manages time as a staff officer and line officer consistently.

Competency: Meets Standards

Comments:

Behavior and Teamwork

- Employee is courteous to the public, administration, and co-workers.

(Captain Core Competencies continued)

- Employee works openly and positively to accomplish city and department goals.
- Employee offers tactful and constructive suggestions.
- When dealing with administration or co-workers, the employee establishes, promotes, and maintains a positive working relationship.

Competency: Meets Standards

Comments:

Emergency Incident Scene

- Employee demonstrates extensive knowledge of Incident Management System.
- Employee uses personnel and equipment to achieve maximum effectiveness, and maintains accountability of personnel under his direct supervision during emergency situations.
- Employee promotes safety practices as defined by LFD rules and regulations.

Competency: Meets Standards

Comments:

Supervisory Skills

- Employee practices and applies the principles of managerial skills and theory when dealing with conflict resolution, discipline, problem solving, and goal setting.
- Employee strives to maintain a work environment free of racial, religious, ethnic, and sexual discrimination and harassment.
- Employee uses written and oral communication effectively, soliciting feedback to determine the level of understanding of the audience.

Competency: Meets Standards

Comments:

Experience Raise *(Value = 50% including written and skill test)*

- Employee satisfactorily completed and passed said experience test.

(Captain Core Competencies continued)

Competency: Meets Standards

Comments:

Signature of Employee	Date
Signature of Supervisor	Date
Signature of Assistant Chief	Date
Signature of Fire Chief	Date

Battalion Chief Core Competencies

Employee Name		Date	02/8/2006
Social Security #		Job Title	Battalion Chief
Time Period Being Evaluated		to	
Department/Division			

Emergency Incident Scene

- Employee demonstrates extensive knowledge of Incident Management System, and effectively applies that knowledge during emergency operations.
- Employee utilizes personnel and equipment to achieve maximum efficiency during emergency situations.

Competency: Meets Standards

Comments:

Officer Development

- Employee assists in the development of officers' managerial skills.
- Employee consistently reviews procedures and improves productivity with new and better work practices.
- Employee develops officers to be effective incident managers.

Competency: Meets Standards

Comments:

Behavior and Teamwork

- Employee projects an image appropriate to the position in terms of dress, hygiene, and behavior.
- Employee follows directions and complies with department and city policies and procedures.
- Employee applies policies and procedures in a manner that is fair and consistent.
- Employee establishes and maintains positive working relationships with outside agencies, other city departments, and the media.
- Employee considers the overall needs of the organization when making decisions or suggestions.

(Battalion Chief Core Competencies continued)

- Employee is flexible when making departmental decisions and when working with other department managers.
- When making suggestions, the employee makes tactful and constructive suggestions.

Competency: Meets Standards

Comments:

Commitment to Public Service

- Employee actively supports city programs, goals, and values within the scope of the position.
- Employee deals with citizens' complaints, concerns, and inquiries in a considerate and timely manner.

Competency: Meets Standards

Comments:

Supervisory Skills

- Employee is innovative in the ability to develop and present new ideas, concepts, and procedures.
- Employee applies policies and procedures in a manner that is fair and consistent.
- Employee organizes work efficiently and effectively.
- Employee delegates responsibilities to his/her officer.

Competency: Meets Standards

Comments:

Administrative Functions

- Employee directs and motivates personnel to produce results.
- Employee works to improve job-related skills and knowledge.
- Employee treats personnel professionally when evaluating performance and applying discipline.
- Employee provides continuing feedback to employees concerning job performance.

(Battalion Chief Core Competencies continued)

Competency: Meets Standards

Comments:

Budget and Area of Responsibility Management

- Employee effectively manages area of responsibility. This includes submitting budget in a timely manner and effectively managing account balances throughout the year. Employee is innovative in maximizing the use of funds in his/her budget.
- Employee ensures officer budget proposals are done correctly and in a timely manner.
- Employee is accessible to officers for budget proposals and review, and assists in prioritizing budget.

Competency: Meets Standards

Comments:

Experience Raise *(Value = 50% including written and skill test)*

- Employee satisfactorily completed and passed said experience test.

Competency: N/A

Comments:

Signature of Employee	Date
Signature of Supervisor	Date
Signature of Assistant Chief	Date
Signature of Fire Chief	Date

Supervisory Core Competencies

Employee Name _____ **Social Security No.** _____
Department/Division _____
Job Title _____ **Date** _____

The following criteria, in conjunction with the *City of Lewisville Policy and Procedure Manual,* formulate the scope of performance in which each City of Lewisville employee must be proficient in order to maintain employment. These core competencies represent the goals and objectives and overall working philosophy of the City of Lewisville. Documentation demonstrating an employee's ability or inability to comply with these criteria, such as commendations or formal disciplinary action, should be attached to the evaluation document.

Accountability

- Accomplishes tasks in the manner and at the time described to others.
- Projects positive image of city to our customers.
- Performs in a manner that builds trust with customers, subordinates, supervisors, and peers.
- While representing the city, maintains a high ethical standard at all times.

_____ Meets conditions for competency.
_____ Does not meet conditions for competency.

Comments: _____
_____

Customer Service/Treats Others Like You Want To Be Treated

- Treats customers in a professional and courteous manner.
- Shows respect for the work of supervisors, subordinates, and peers.
- Identifies, understands and responds to appropriate needs of customers.
- Responds appropriately to *both* internal and external customers.

_____ Meets conditions for competency.
_____ Does not meet conditions for competency.

Comments: _____
_____

(Supervisory Core Competencies continued)

Teamwork

- Support city/department goals and objectives.
- Puts interest of city/department ahead of self.
- Recognizes and respects the contributions and needs of all individuals.
- Builds and maintains productive working relationships.
- Generates and promotes ideas and implements actions which build upon teamwork and encourages others to do so.

_____ Meets conditions for competency.
_____ Does not meet conditions for competency.

Comments: _____

Policy Compliance

- Observes safety rules and procedures.
- Strict adherence to assigned work schedules, leave policies and attendance guidelines.
- Complies with all division/department policies and procedures.

_____ Meets conditions for competency.
_____ Does not meet conditions for competency.

Comments: _____

Effective Communication

- Clearly communicates with *both* internal and external customers in a timely, accurate manner.
- Communicates in an appropriate, courteous manner commensurate with job duties.
- Discusses concerns and conflicts, seeking a constructive and timely resolution.
- Utilizes proper phone etiquette.
- Keeps the public, associates, and supervisors informed.
- Listens effectively.

_____ Meets conditions for competency.
_____ Does not meet conditions for competency.

(Supervisory Core Competencies continued)

Comments: _____

_____

Resource Management

- Makes knowledgeable, timely decisions.
- Follows through to deliver results.
- Uses resources such as time, materials, equipment, and supplies in the most effective manner.
- Strives for ways to increase productivity and reduce cost.

_____ Meets conditions for competency.
_____ Does not meet conditions for competency.

Comments: _____

_____

Leadership (Applicable to supervisory personnel only)

- Keeps employees and citizens informed of decisions and actions.
- Accepts responsibility and accountability for work group results.
- Develops teamwork by managing, planning, and coordinating group efforts to achieve work group goals.
- Encourages employees to contribute to problem identification, solution development, and implementation.
- Conducts quality employee performance evaluation following established guidelines and timelines.
- Acts on the basis that change is essential to progress.
- Openly asks for constructive criticism and suggestions from customers and employees.
- Ensures that employees both understand *and* comply with city/department policies and procedures.

_____ Meets conditions for competency.
_____ Does not meet conditions for competency.

Comments: _____

_____

SUMMARY:

_____ **Met all conditions of competency on Core Competency Evaluation.**
_____ **Did not meet conditions of competency on one or more factors.**
(Employee has been placed on a Performance Improvement Plan.)

_____   _____
Employee Signature   Date

Non-Supervisory Core Competencies

Employee Name _____ **Social Security No.** _____
Department/Division _____
Job Title _____ **Date** _____

The following criteria, in conjunction with the *City of Lewisville Policy and Procedure Manual*, formulate the scope of performance in which each City of Lewisville employee must be proficient in order to maintain employment. These core competencies represent the goals and objectives and overall working philosophy of the City of Lewisville. All documentation demonstrating an employee's ability or inability to comply with these criteria, such as commendations or disciplinary actions, must be attached to the evaluation document.

Accountability

- Accomplishes tasks in the manner and at the time described to others.
- Projects positive image of city to our customers.
- Performs in a manner that builds trust with customers, subordinates, supervisors, and peers.
- While representing the city, maintains a high ethical standard at all times.

_____ Meets conditions for competency.
_____ Does not meet conditions for competency.

Comments: _____
_____

Customer Service/Treats Others Like You Want To Be Treated

- Treats customers in a professional and courteous manner.
- Shows respect for the work of supervisors, subordinates, and peers.
- Identifies, understands and responds to appropriate needs of customers.
- Responds appropriately to *both* internal and external customers.

_____ Meets conditions for competency.
_____ Does not meet conditions for competency.

Comments: _____
_____

(Non-Supervisory Core Competencies continued)

Teamwork

- Support city/department goals and objectives.
- Puts interest of city/department ahead of self.
- Recognizes and respects the contributions and needs of all individuals.
- Builds and maintains productive working relationships.
- Generates and promotes ideas and implements actions which build upon teamwork and encourages others to do so.

_____ Meets conditions for competency.
_____ Does not meet conditions for competency.

Comments: _____
_____

Policy Compliance

- Observes safety rules and procedures.
- Strict adherence to assigned work schedules, leave policies and attendance guidelines.
- Complies with all division/department policies and procedures.

_____ Meets conditions for competency.
_____ Does not meet conditions for competency.

Comments: _____
_____

Effective Communication

- Clearly communicates with *both* internal and external customers in a timely, accurate manner.
- Communicates in an appropriate, courteous manner commensurate with job duties.
- Discusses concerns and conflicts, seeking a constructive and timely resolution.
- Utilizes proper phone etiquette.
- Keeps the public, associates, and supervisors informed.
- Listens effectively.

_____ Meets conditions for competency.
_____ Does not meet conditions for competency.

(Non-Supervisory Core Competencies continued)

Comments: _____

_____

Resource Management

- Makes knowledgeable, timely decisions.
- Follows through to deliver results.
- Uses resources such as time, materials, equipment, and supplies, in the most effective manner.
- Strives for ways to increase productivity and reduce cost.

_____ Meets conditions for competency.
_____ Does not meet conditions for competency.

Comments: _____

_____

SUMMARY:

____ **Met all conditions of competency on Core Competency Evaluation.**

____ **Did not meet conditions of competency on one or more factors.**
(Employee has been placed on a Performance Improvement Plan.)

_____   _____
Employee Signature   Date

_____   _____
Supervisor Signature   Date

Self-Evaluation given to employee_____
Date

Supervisor's Initials_____

EMPLOYEE SELF ASSESSMENT

EMPLOYEE:

JOB TITLE:

REVIEW PERIOD FROM:

NAME OF EVALUATOR:

(Above to be completed by supervisor prior to giving to the employee)

If you would like assistance in completing this form, contact your supervisor or the Human Resources Department.

_____
EMPLOYEE SIGNATURE

GIVE THIS FORM TO YOUR SUPERVISOR NO LATER THAN TWO WEEKS AFTER THE DAY YOU RECEIVED IT

****

INSTRUCTIONS:

An important part of the performance appraisal process is the opportunity for employee involvement. The Employee Self Assessment is provided to solicit information from employees about performance. This is an opportunity for you to give the evaluator your comments and conclusions about performance. It is an opportunity for you to communicate with your supervisor concerning any areas of concern about your job. Please take some time to think about the questions before answering them.

****

(IF ADDITIONAL SPACE IS NEEDED, ATTACH ADDITIONAL PAGES)

(Employee Self Assessment continued)

HOW WELL DID I DO IN MY JOB DURING THE REVIEW PERIOD?

WHAT DO I LIKE BEST ABOUT MY JOB?

WHAT DO I LIKE LEAST ABOUT MY JOB?

WHAT WERE MY ACCOMPLISHMENTS DURING THE PERIOD?

WHAT CAN I DO TO BE BETTER AT MY JOB?

WHAT COULD THE CITY/MY SUPERVISOR DO WHICH WOULD HELP ME DO MY JOB BETTER?

PROGRESS IN ATTAINING PREVIOUSLY SET GOALS:

GOAL 1

EMPLOYEE COMMENTS:

GOAL 2.

EMPLOYEE COMMENTS:

GOAL 3.

EMPLOYEE COMMENTS:

GOAL 4.

EMPLOYEE COMMENTS:

GOAL 5.

EMPLOYEE COMMENTS:

(Employee Self Assessment continued)

GOALS SET BY EMPLOYEE FOR NEXT EVALUATION:

Use this section to set goals and objectives you would like to meet during the next review period. Goals should be challenging, measurable, and attainable. Next, give this completed form to your supervisor. When you meet to review your performance evaluation form and self assessment form, together you will mutually establish goals for the next review period.

1. _____
_____
_____
_____

2. _____
_____
_____
_____

3. _____
_____
_____
_____

4. _____
_____
_____
_____

5. _____
_____
_____
_____

Employee Signature _____ Date _____

Supervisor Signature _____ Date _____

Key Results for Division Managers

Performance Management Plan
Department Director and Division Manager

Name	Job Title	Review Period

Fire		Administration
Department		**Division**

Key Results:

Key results are defined as expectations of the city manager for a department director *or* a department director for a division manager in a defined review period. These key results are focused on what the department director/division manager should achieve through use of available resources including personnel, equipment, etc. and not necessarily what the person achieves independently. Key results are future-oriented and should take into account key issues and not day-to-day operational issues. Key results must be flexible and may be changed throughout the performance period as requirements and conditions change.

Key Result #1
_____
_____
_____

Key Result #2
_____
_____
_____

Key Result #3
_____
_____
_____

Key Result #4
_____
_____
_____

(Key Results continued)

Key Result #5
_____
_____
_____

Key Result #6
_____
_____
_____

Key Result Changes During Review Period:
_____
_____
_____
_____

Key Results: Evaluation of Prior Year Results

Key Result #1
_____
_____
_____

Key Result #2
_____
_____
_____

Key Result #3
_____
_____
_____

Key Result #4
_____
_____
_____

Key Result #5
_____
_____
_____

(Key Results continued)

Key Result #6
_____
_____
_____

Executive Competencies:

Competencies are defined as demonstrable characteristics of the person, including knowledge, skills, and behaviors that enable successful performance. Competencies add value by communicating what the person must know to help the organization succeed.

(1)   Team/Co-Worker Effectiveness Orientation

- Places a premium on contributing to the success of co-workers.

- Serves as effective representative of the department/division by directing activities toward the joint attainment of objectives with other departments/divisions as well as with the city council, outside agencies, and Lewisville citizens.

- Recognizes that "Team" includes all departments/divisions in the organization and establishes priorities that are most beneficial to the "Team" and not just to one department/division. Does not base decisions on protecting one's "turf."

_____ Demonstrates competency
_____ Does not demonstrate competency

_____
_____
_____

(2)   Interpersonal and Communication Effectiveness

- Demonstrates skill in multidirectional communication rather than merely top-down communication to assure that employees and citizens have information needed.

- Focus on extensive sharing of information throughout the department/division using a variety of communication tools.

- Demonstrates a "People-Involved" method of communication rather than a "People-Told" method.

- Conveys thoughts clearly and concisely, both verbally and in writing.

- Shows empathy—understands and appropriately considers the needs and problems of others.

_____ Demonstrates competency
_____ Does not demonstrate competency

_____
_____
_____

(Key Results continued)

(3) <u>Reliability and Results Orientation</u>

- Prioritizes effectively; focuses on the most important issues.
- Delivers what is promised to others by the date agreed.
- Ensures that staff activities lead to productive results.
- Takes appropriate action in a timely manner rather than studying an issue for an overly-long period of time.

_____ Demonstrates competency
_____ Does not demonstrate competency

_____
_____
_____

(4) <u>Leadership Effectiveness</u>

- Demonstrates respect for the individual through honesty and following the golden rule.
- Holds others accountable for value-driven performance and behavior.
- Shows sincere appreciation for employee contributions and achievements and provides recognition through Quality Award Program.
- Establishes measurable standards for subordinate positions which are agreed to and align horizontally and vertically within the organization.
- Establishes stretch goals and increasingly higher standards for subordinates.
- Treats employees with respect and encourages an open exchange of ideas for improvement.
- Demonstrates daily behavior that aligns with the city manager's operating value statements and the City of Lewisville mission.
- Focuses on decision-making based on collaboration and consensus-building rather than unilateral action.
- Contributes to effective resolution of complaints/disagreements.

_____ Demonstrates competency
_____ Does not demonstrate competency

_____
_____
_____

(Key Results continued)

(5) <u>Customer-Focus Orientation</u>

- Demonstrates a recognition that customers are both external and internal to the organization and that customer needs should be identified and responded to in a timely manner.

- Requires subordinates to make superior customer service a top priority.

_____ Demonstrates competency
_____ Does not demonstrate competency

---

(6) <u>Adaptability and Innovation Effectiveness</u>

- Promotes and initiates useful change.

- Encourages others to improve operational and administrative processes and procedures.

- Demonstrates creative problem-solving.

- Demonstrates tolerance for honest mistakes/views failure as a growth opportunity.

- Takes prudent risk to increase effectiveness rather than demonstrating risk avoidance.

- Regularly questions "the way we do things now."

_____ Demonstrates competency
_____ Does not demonstrate competency

---

(7) <u>System Effectiveness</u>

- Supports organizational policies and procedures.

- Properly applies policies and procedures including policies related to performance management, employee discipline, drug testing, financial management, and purchasing processes.

_____ Demonstrates competency
_____ Does not demonstrate competency

(Key Results continued)
Summary:

Overall Evaluation:

_____   Key results achieved and executive competencies obtained.

- Merit Adjustment _____

_____   Some key results not obtained or some competencies not fully demonstrated.

- Merit Adjustment _____

_____   Either key results not achieved or executive competencies not obtained.

- Results in merit increase being withheld for a one year period.
- Employee placed on Improvement Plan for 90 days to one year period.
- Consideration of termination at end of Improvement Plan if required changes not obtained.

Employee Comments:
_____
_____
_____
_____

Supervisor Comments:
_____
_____
_____
_____

_____   _____
Employee Signature   Date

_____   _____
Supervisor Signature   Date

Employee Work Plan

                                    Annual_____ **Three Month**_____

Employee Name _____ **Social Security No.** _____
Department/Division _____
Job Title _____ **Date** _____

In the space provided below, list the five top priorities, goals/tasks, and standards for the upcoming review period. For each goal, identify the internal or external customers served and quality level (measurable indicators) that will lead to accomplishment of the City's and Department's mission, goals and objectives. Results and comments should be made during the next scheduled performance review. Work plans may be amended during the evaluation period.

1.   A.   Goal/Task:

       B.   Customer:

       C.   Standards required (measurable indicators):

       D.   Results/Comments (to be completed next evaluation period):

2.   A.   Goal/Task:

       B.   Customer:

(Employee Workplan continued)

        C.   Standards required (measurable indicators):

        D.   Results/Comments (to be completed next evaluation period):

3.   A.   Goal/Task:

        B.   Customer:

        C.   Standards required (measurable indicators):

        D.   Results/Comments (to be completed next evaluation period):

4.   A.   Goal/Task:

        B.   Customer:

        C.   Standards required (measurable indicators):

        D.   Results/Comments (to be completed next evaluation period):

(Employee Workplan continued)

5.   A.   Goal/Task:

        B.   Customer:

        C.   Standards required (measurable indicators):

        D.   Results/Comments (to be completed next evaluation period):

I have read and discussed this work plan with my supervisor and accept this plan as written.

_____   _____

Employee Signature   Date

(Employee Workplan continued)

WORK PLAN SUMMARY

I.   Type of check

_____   3 month (new hires/promotions)

_____   Annual

II.   Improvements/accomplishments over past period:

III.   Obstacles (If employee did not accomplish goals/performance tasks, list reasons here):

   A.   Within employee's control

   B.   Outside employee's control

IV.   Employee comments

V.   Overall assessment of employee's performance on work plan:

*   ___ (a)   Did not meet quality standard on one or more goals/tasks on work plan.

   ___ (b)   Did not meet all quality standards on work plan but obstacles existed.

   ___ (c)   Met all quality standards on work plan.

*   A checkmark on (a) will result in the employee being placed on an Improvement Plan and the raise withheld for the specified time period.

_____   _____
Signature of employee   Date

_____   _____
Signature of supervisor   Date

(Employee Workplan continued)

Exempt Employees/Exceptional Performance Pay

(Excludes Police and Fire Rank Employees)

Exceptional performance pay is based on accomplishments that significantly impact the city's/department's mission and goals. The criteria for receiving exceptional performance pay include accomplishments that are <u>self-motivated</u> and involve (1) significant budget savings, (2) the use of new technology or an improved process that results in improved customer service or significant budget savings, or (3) initiating projects with successful results.

In order to receive exceptional performance pay, the evaluating supervisor must submit a description of the performance to the department director prior to any discussions with the employee. The department director will note whether or not the description meets the required criteria. The Human Resources Department will submit an annual report to the city manager of all exceptional performance pay that is distributed. All approved descriptions for exceptional performance pay should be attached to this report.

Exceptional Performance Pay ____ Yes

____ No

_____ _____
Signature of Employee Date

_____ _____
Signature of Supervisor Date

B

DEPARTMENTAL COMMUNICATIONS

Our Foundation

Our Community

★ **Our City's Mission** ★

To enhance the quality of life for our community and provide effective municipal services

★ **Our City's Vision** ★

Unified community committed to excellence

★ **Our City's Values** ★

- ★ Communication
- ★ Competence
- ★ Customer Satisfaction
- ★ Efficiency
- ★ Enthusiasm
- ★ Innovation
- ★ Integrity
- ★ Teamwork
- ★ Trust

★ OUR MISSION ★

 BE HONEST ★ BE SAFE ★ BE NICE
TREAT PEOPLE LIKE FAMILY

★ OUR VISION ★

TO PROVIDE THE BEST SERVICE
AND PROTECTION IN THE COUNTRY

 ★ OUR VALUES AND GUIDING PRINCIPLES ★
PRIDE, HONOR AND INTEGRITY
~~ OUR FOUNDATION ~~

TRAINING / SAFETY DIVISION

»» *PASS IT ON PROGRAM* »»
12.04.1

*** SAFETY BULLETIN ***

Car Hood Strut Explosion

Overview

On Sunday November, 21, 2004, the Windsor Locks Fire Department responded to a car fire. Upon arrival the fire was contained to the engine compartment. While gaining access to the engine, one firefighter was releasing the hood's safety latch when the hood shock strut exploded and fired into the firefighter, striking him in his upper thigh, completely piercing his bunker pants and his leg. The strut was approximately 18 inches long. The struts are filled with a compressed gas and are common on GM cars, in particular Buicks.

Please beware of these and other gas filled cylinders (bumper shocks, drive trains, etc.)

The firefighter is home recuperating. The extent of his injury and ability to return to work is not yet known.

BE SAFE

LFD Newsletter

First Due

City of Lewisville Fire Department

3rd **Quarter**
2005

Preparing for Cold Weather Operations. . .

It is getting close to that time of year when Old Man Winter sets in and temperatures drop below freezing. So we need to get the Fire Houses ready. There are the usual considerations that you would think about around your own homes, such as wrapping outside faucets to prevent freezing, disconnecting hoses from the faucets, and replacing return air filters so that the heaters operate more efficiently. It is also time to check bay heaters to make sure they are working properly and report any needed repairs so that they can be fixed before the heaters are put in use. This is also the time of year that the bay door timers need to be left on, this way you can be certain that the doors will close after you leave on calls and the bay will be warm when you return. It is important to realize that the heaters in the bay are to keep temperatures around 58 degrees, so the lines on the apparatus do not freeze.

Also, let's talk about your own personal welfare. It is important that you remember to stay hydrated. I know you aren't as thirsty around the Fire House when it is cold, but you still need water. It is important to stay hydrated so your body will operate at its optimum level when you go on that middle of the night fire. It is also important that you dress appropriately for the cold weather, don't over dress if you are putting on your bunker gear to combat a fire. If you overdress you will sweat more, and this can cause you to get chilled quicker under your PPE.

Finally, and most importantly, let's talk about your apparatus. Without these working properly you can't do your job. Things to remember are; engage your pump when sitting at a scene and not pumping a fire, let the water warm a bit and then turn on the circulator so that the warm water will keep the lines from freezing. Also it is important to keep your drain valve open when not in use. This keeps excess water from building in the lines and freezing. You can also wipe anti-freeze around the threads of your discharges and hose connections. This will ensure that when you need to take a cap off or disconnect a hose you will be able to do so without fear of ice having formed and making a disconnect hard to accomplish.

Our Public Works Department is great at sanding our ramps when ice starts to form on the City's roadways. Just remember they also have to prioritize and they may need to do bridges and overpasses before they get to us, be patient. It is also important to make sure all apparatus' tire chains are located and in good condition before you need them. If you find a problem with your chains, let Chief Covey know so that he can replace them.

Even though I enjoy a good cold spell I hope we don't encounter too harsh a winter in Texas. But if we do, we can be sure to operate at our best if we take a little time before hand to make sure things are working properly and that our apparatus and ourselves are ready to go.

Tim Tittle
Assistant Chief
Operations

Inside this **issue:**

The Watch Desk	2
Historically Speaking	2
Statistics	3
From the Flight Deck	4
Word Search	4
Auxiliary Cooling Devices	4
. . .And Some Do It For Free!	5
The Probie (aka New Hire)	5
Introducing	6
So You Want to Listen to FDNY	7
FY 05/06 Budget	7
LAFS Clown News	8
LFD Honor Guard	8
Department News	9, 10

The Life of a Fire Chaplain. . .

Since 911, the position of Fire Chaplain has gained prominence. Little did I realize when I became a Fire Chaplain 11 years ago what a change it would bring to this ministry. Make no mistake, it truly is a ministry, and it had better be a calling from God.

I became the Fire Department Chaplain when the former Chaplain asked me to fill in for him one Sunday. Since then, I have had the privilege of ministering every Sunday morning to the firefighters I also minister to their families and the community during the darkest hours of their lives.

When the phone rings or you're paged, you have to be ready, for you are on call 24/7. You must be prepared to minister to people in the greatest crisis of their lives, Such as standing with a family when everything they own has gone up in smoke, a firefighter who was injured fighting a fire, delivering a death message, standing with parents when a child has died, holding children when they realize the loss of a loved one or just ministering and loving people when they face devastating problems.

After particularly difficult calls, it is my duty to check on the firefighters, the paramedics, the dispatchers, even the doctors and nurses to offer diffusing or simply visit with them over a cup of coffee. We often don't realize just how deeply their lives are effected. I may not say a lot during these times, because in a time of crisis it is not who I am nor what I say, but who I represent.

How can a chaplain make a difference in people's lives? By being there for them, loving them as Christ would love. The benefits come. . .when a dad calls and says, "chaplain, thanks for being there last night, just wanted to tell you God is working in our lives", or you receive a card expressing thanks, or nurses from the hospital bring you cookies, or just the fact that you get to minister and be a part of the Fire Department family. The final benefit will be to hear the one that I represent say, "Well done."

Bennie Grissom
Chaplain

Page 2 — First Due

The Watch Desk...

Richard Lasky
Fire Chief

Good procedures are so simple you don't need to write them down to remember them or use a dictionary to understand them

For those that may have never heard of the term the "Watch Desk" and there may be a few, it dates back to when it all started, back in the horse and steam engine days. The watch desk was usually located near the front of the firehouse and was staffed by a firefighter 24/7. This was where assignments for fires and other calls for help came through, usually by way of telegraph (that whole Box Alarm thing) and later by radio speaker and then by phone. This was the place where both department and personal phone calls were received and "paged out" and was the source of information for pretty much anything going on, whether it was a fire in another part of town, a drill, something happening at headquarters and a whole list of other bits and pieces of information. It was looked upon as the best and most efficient means for communicating information and getting answers to questions. The firefighter manning the watch desk for whatever period of time during a shift was always in the know. In some places around the country such as in the FDNY it still exists and holds true to its original concept.

And even though we don't have a watch desk with the exception of the captain's office, are you still staying informed? Are you in the know? Are you tuned in to what's going on? We can still have that firehouse watch desk... ...if you want to.

And lastly, below is something that's been around a while and with as many new members as we have coming on board, it might not be a bad time to take a look at again.

Be safe!

Firefighters Ten Commandments

1. Be sincerely interested in, and dedicated to, your job. What you are able to contribute to and receive from, your department is only limited by your own degree of personal commitment.
2. Be loyal to your department and to your coworkers. You are a part of your department, and it is a part of you.
3. Be diligent to know and like your job. Those who never do more than they are paid to do never get paid more for what they do. Do your best to analyze and profit by your mistakes. Be conscientious, recognize and accept your responsibilities, including responsibilities for which you are not specifically assigned.
4. Be aggressive in the pursuit of all education and training opportunities. You are never fully trained. The achievement of each educational objective only reveals the path to the ever increasing body of knowledge with which a firefighter must be familiar if he is to excel in this challenging profession.
5. Be courteous, considerate, enthusiastic, and cooperative. You were considered to be this kind of person when you were hired. Be especially tactful and considerate in dealing with those who have experienced a loss due to fire.
6. Be constantly aware that you are a representative of the Fire Department. Be certain that your dress and actions are a credit to this honorable profession.
7. Be cautious, guard your speech, both on and off duty. As a member of the Fire Department, it is expected that you may possess information which should not be revealed. Handle privileged communications as such, but be always willing to discuss the purpose, functions, history, and traditions of the Fire Service.
8. Be the kind of person who inspires confidence and respect. Do this by being honest, fair, and trustworthy in all your dealings with others, and by keeping your personal affairs in such order that they will never embarrass you or your department if made public.
9. Be able to accept criticism graciously, and praise, honor, and advancement modestly. Be aware of the fact that your personality is never completely developed. You have an obligation to all with whom you interact to continually try to grow.
10. If any would be great among us, let him first learn to serve.

By Carl E McCoy

Historically Speaking...

The badge of a firefighter is the Maltese Cross. The Maltese Cross is a symbol of protection, a badge of honor, and its story is hundreds of years old. When a courageous band of crusaders, known as the Knights of St. John, fought the Saracens for possession of the Holy Land, they were faced with a new weapon known to European fighters. It was a simple but horrible device of war. The Saracens weapon was fire.

As the crusaders advanced on the walls of the city, they were saturated with glass bombs containing naptha. When they were saturated with the liquid, the Saracens threw flaming torches into the crusaders. Hundreds of knights were burned alive while others risked their lives in an effort to save their kinsmen from painful fiery deaths. Thus these men became the first firemen, and the first of a long line of Firefighters. Their heroic efforts were recognized by fellow crusaders who awarded each other with a badge of honor similar to the cross firefighters wear today.

Since the Knights of St. John lived on the island of Malta in the Mediterranean Sea for close to four centuries, the cross came to be known as the Maltese Cross. The Maltese Cross is your symbol of protection. It means that the Firefighter that wears this cross is willing to lay down his life for you, just as the crusaders sacrificed their lives for their fellow man so many years ago. The Maltese Cross is a Firefighter's badge of honor, signifying that he works in courage—a ladder rung away from death

| 3rd Quarter | Page 3 |

Call Statistics...

YTD 6,289 Calls

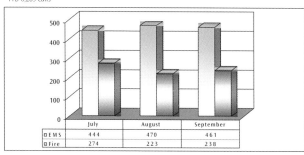

	July	August	September
EMS	444	470	461
Fire	274	223	238

A special thank you to the nocturnal truck on C-shift for helping us pick up the slack in the wee hours of the morning.

Training Statistics...

YTD 22,591.5 Training Hours

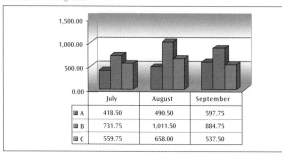

	July	August	September
A	418.50	490.50	597.75
B	731.75	1,011.50	884.75
C	559.75	658.00	537.50

Train as if your life depends upon it, because it does.

Ambulance Collection Stats...

4th Quarter
FY 04/05 Collections: $834,990.83

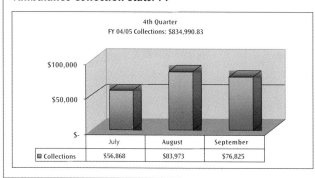

	July	August	September
Collections	$56,868	$83,973	$76,825

Thanks to all 3 shifts for good documentation and hats off to Mary Ann for a great job.

Last Year Collections:
$744,506.00

Weekly Update
BURNING ISSUES

...with Pride, Honor and Integrity

Week of: July 3, 2005 to July 10, 2005

OPERATIONS DIVISION

SPECIAL INTEREST CALLS

Major Accident

On July 8, 2005 at 9:30 PM, A-Shift, Quint 162 responded to a major accident between a car and a motorcycle. Careflite was requested and the Battalion Chief was dispatched to set up a landing zone. The accident site was blocked off by the Police Department for traffic control. The driver of the car refused transport.

Thermal Imaging Camera

On Saturday, July 9, 2005, B-Shift Truck 161 assisted the Police Department in searching for some suspects by using one of our thermal imaging cameras.

Structure Fire

On July 10, 2005, C-Shift responded to a structure fire at 922 Sylvan Creek. They arrived on scene to find moderate smoke from the roof vents over the garage. They cut a ventilation hole over the garage. The fire was extinguished by using approximately 20 gallons of water. Engine 161 performed a secondary search. The cause of the fire was undetermined and the investigation is on-going. The house was checked for rekindle approximately two hours later.

WEEKLY STATISTICS

July 4, 2005 through July 10, 2005

EMS Calls	101
Fire Calls	47
Total	148
Year-To-Date Total Calls	4059

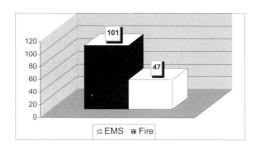

SUPPORT SERVICES

TRAINING

Training for month of June, 2005   1,764.50 hours

Year to date total   12,130.50 hours

Multi-Agency Night Drills

The Lewisville, Highland Village, Flower Mound and Lake Cities multi-agency training on July 5, 6 & 7, 2005 went well. The evolution was a Two Story Commercial Fire. Everyone in our Department took part.

Dive Team Swim Tests

The Dive Team has completed its baseline swim tests at Sun Valley Pool. The Divers were required to swim 500 yards freestyle; tread water for 15 minutes, the last two with their hands above their head; an 800 yard snorkel and a diver rescue pull of 100 yards. Their scores will be computed and they will be required to train in any area of weakness. The tests will be given annually beginning in August of 2006.

FIRE SIDE

FIREFIGHTER OLYMPICS

The 2005 Firefighter Olympics are scheduled to begin on Sunday, July 17 and run through Friday, July 22. There are many events and programs scheduled during the week. Be sure to come out to the various venues and root for the "home team".

CEREMONIES

Effective Date:_____
Approval:_____

AWARDS DESCRIPTIONS

STANDARD OPERATING PROCEDURES

SECTION: PERSONNEL
TOPIC: AWARDS, DECORATIONS AND COMMENDATIONS
REFERENCE: 17.02

I. AWARDS, DECORATIONS AND COMMENDATIONS

A. The Lewisville Fire Department has established an awards program to provide a method whereby deserving employees, civilians, members and officers, as well as the general public, will receive official departmental and public recognition for outstanding acts of valor or meritorious service, on or off duty. Proper placement of resulting decorations on uniforms is included.

B. All Lewisville Fire Department personnel are encouraged to be alert for individuals deserving of recognition and follow through with making nominations to the committee.

C. Procedure for Nomination:

1. Obtain form from Lewisville Fire Department secretary
2. Fill out form
3. Submit form to chief of department who will pass it on to the committee chairman
4. Chairman assigns form to committee member for investigation
5. Committee member reports findings
6. Committee votes, majority rules
7. Award sanctioned or denied
8. Award presented at ceremony

D. Awards Committee
   1. Consists of two representatives from each shift, both of whom can be a combination of any rank from firefighter to battalion chief
   2. Elected by majority vote of members/officers from each shift for a term of two years
   3. Chairman will be assistant chief of operations
   4. Committee meets biannually, 45–60 days prior to ceremony
   5. Committee accepts and votes on nominations, majority rules
   6. Committee will make a timely acknowledgement of all nominations received
   7. Assigned committee member will investigate nominations and report results during biannual meeting
   8. Committee will purchase all awards, certificates, etc.
   9. Committee will design and keep a revised nomination form; this form is to be available from Lewisville Fire Department secretary
E. Awards for Lewisville Fire Department members
   1. The Lewisville Fire Department Award of Valor
      a. The highest award presented by the department
      b. Awarded only to Lewisville Fire Department members and officers
      c. Available for both on-duty and off-duty acts
      d. Criteria:
         1) Any time a member/officer of the Lewisville Fire Department is killed in action while involved in emergency operations
         2) Any time a member/officer who, by such intrepid display of gallantry, distinguishes him/herself in a conspicuously and intelligent manner, and at the risk of his/her life. The deed performed must have been by voluntary act of personal bravery and intrepidity above his/her comrades, and must have involved risk of life, known to the individual while performing the act.
   2. The Lewisville Fire Department Award of Merit
      a. The second-highest award presented by the Lewisville Fire Department
      b. Awarded only to Lewisville Fire Department members/officers
      c. Criteria:
         1) Any time a member/officer has performed a courageous act under conditions of extreme danger and at the exceptional risk of serious bodily injury that clearly exceeds what is normally required, which has resulted in the protection of life and is considered to be exceptionally valorous
         2) Serious bodily injury is defined as any injury causing death or requiring hospitalization and/or two 24-hour shifts off.

3. The Lewisville Fire Department Company Citation
   a. Criteria:
      1) Awarded to a unit whose performance was an unequaled team effort in overcoming unusual difficulties or obstacles in the completion of a difficult task that has resulted in the protection of life or property
      2) A unit is defined as any company or operational/functional group consisting of at least two members/officers.
4. The Lewisville Fire Department Award of Commendation
   a. Awarded to Lewisville Fire Department members/officers and civilian employees
   b. Criteria:
      1) Awarded for commendable services rendered over an extended period of time resulting in raising the standards of the department or fire service profession
      2) Awarded for commendable service rendered during emergency operations above and beyond the normally expected, resulting in the protection of life or property, but not qualifying for the Lewisville Fire Department Award of Valor or Merit
5. The Lewisville Fire Department Firefighter of the Year Award
   a. Criteria:
      1) To entitle a member for consideration of this award, he/she is described as "The type of person you call upon to get the job done completely, competently and correctly. He/she works well with everyone and is willing to learn as well as to teach his/her peers. He/she strives for job excellence and will put in extra hours to assure the department will reach its goals. He/she is a positive role model."
6. Lewisville Fire Department Paramedic of the Year
   a. Criteria:
      1) To entitle a paramedic for consideration of this award, he/she is described as "The type of person you call upon to get the job done completely, competently and correctly. He/she works well with everyone and is willing to learn as well as teach his/her peers. He/she strives for job excellence and will put in extra hours to assure the department will reach its goals. He/she is a positive role model."
7. Lewisville Fire Department Officer of the Year
   a. Criteria:
      1) To entitle a member for consideration of this award, he/she is described as "The type of person you call upon to get the job done completely, competently and correctly. He/she works well with everyone and is willing to learn as well as teach his/her peers. He/she strives for job excellence and will put in extra hours to assure the department will reach its goals. He/she is a positive role model."

9. Lewisville Fire Department Rookie of the Year
    a. Firefighters with less than two years service with the Lewisville Fire Department
    b. Criteria:
        1) To entitle a member for consideration of this award, he/she is described as "The type of person you call upon to get the job done completely, competently and correctly. He/she works well with everyone and is willing to learn as well as teach his/her peers. He/she strives for job excellence and will put in extra hours to assure the department will reach its goals. He/she is a positive role model."

F. Civilian Awards
    1. Award of exemplary action (civilians, police officers, members of other fire departments, etc.)
        a. To entitle a civilian for consideration of this award, the act must have been of unusual character requiring initiative or ability worthy of special recognition.
        b. His or her actions must have resulted in a positive impact to life and/or property.
        c. The recommendation may come from a member of the department stating the reasons for the award, and be directed to the fire chief
        d. The committee will consider the nominations as presented by the fire chief.
    2. Lewisville Fire Department Letter of Appreciation
        a. To entitle a civilian for consideration of this letter, his or her actions must warrant recognition by the fire department.
        b. The recommendation may come from a member of the department stating the reasons for the award and be directed to the fire chief.
        c. The committee will consider the nominations as presented by the fire chief.

G. Presentation of Awards
    1. Medal of Valor
        a. A letter from the chief of the department briefly outlining the circumstances under which the award was earned, along with comments
        b. Medal and shirt bar
        c. Certificate suitable for framing
    2. Awards of Merit and Commendation
        a. A plaque with the member's name which briefly outlines the circumstances warranting the award
        b. Shirt bar
        c. Certificate suitable for framing

3. Unit Citation
    a. A letter from the chief of the department briefly outlining the circumstances under which the award was earned along with comments
    b. A plaque with the members'/officers' names which briefly outlines the circumstances warranting the citation, to be displayed in the station or wherever appropriate
    c. Shirt bar
    d. Certificate suitable for framing
4. Letter of Appreciation
    a. A Letter from the chief of the department briefly outlining the circumstances under which the letter is warranted
5. Firefighter/Paramedic of the Year, Officer of the Year, Rookie of the Year
    a. A Letter from the chief of the department briefly outlining the reasons the individual was chosen
    b. Certificate suitable for framing
    c. Shirt bar
    d. Master plaque to be displayed in the department's administrative office in an area that displays all previous awards and the year they were awarded
6. All awards will be presented at an appropriate public function with the exception of the letter of appreciation, which will be presented as it is earned.
    a. Efforts will be made in advance for publicity and good media coverage.
    b. A record of all awards will be made and kept as a permanent part of each recipient's personnel file.
    c. Awards for prior service with another fire department may be authorized for wear by written approval of the fire chief.

H. Awards and Decorations
   1. Award of Valor
       a. Color bar: red/white/red
       b. Ribbon
           1) Color: red/white/red
           2) Silver/gold medallion
       c. Subsequent awards indicated by oak leaf cluster
   2. Award of Merit
       a. Color ribbon: blue/white/blue
       b. Silver/gold medallion

3. Award of Commendation
    a. Color ribbon: green/white/green
4. Company Citation
    a. Color ribbon: red/white/blue/white/red
5. Firefighter of the Year
    a. Red bar
6. Paramedic of the Year
    a. Blue bar
7. Officer of the Year
    a. White bar
8. Rookie of the Year
    a. Green bar
9. Ribbons: dress uniform, all ranks, all personnel
    a. Name badge is centered on top of the coat's left coat seam.
    b. The "9-11" badge shall be centered above the name badge unless the member has additional ribbons.
    c. The ribbons are worn with the highest award ribbon next to the heart.
    d. Order of importance:
        1) Valor
        2) Merit
        3) Commendation
        4) Company citation
        5) Officer of the Year
        6) Paramedic of the Year
        7) Firefighter of the Year
        8) Rookie of the Year

II. INSPECTIONS

A. Each member of the fire department will have all uniform components inspected twice a year; one in or near the month of January and the other in late summer, due to budget projections and fiscal year completion of the current budget.

B. The quarter master will determine the inspection dates and appoint personnel to conduct the yearly inspection.

C. Items that fail the inspection will be replaced as soon as possible.

Awards Form

Recommendation for Department Award
Lewisville Fire Department

☐ Award of Valor
☐ Award of Merit
☐ Award of Commendation
☐ Company Citation
☐ Award of Exemplary Action

☐ Letter of Appreciation
☐ Firefighter of the Year
☐ Paramedic of the Year
☐ Officer/Employee of the Year
☐ Rookie of the Year

Check Appropriate Box(es)

Name of Member Recommending Award	Rank	Assignment
Name of Nominee	Rank	Assignment
Company/Unit(s) (For Company Citation Only)	Incident	Date of Incident

FACTS SUPPORTING AWARD RECOMMENDATION (For Company Citation list members being cited)

_____
_____
_____
_____

Signature, Member Recommending Award	Date Submitted	☐ Facts Continued on Reverse Side

Service/Awards Committee Signatures

☐ Approve   Signature & Title   ☐ Approve   Signature & Title
☐ Disapprove   ☐ Disapprove

☐ Approve   Signature & Title   ☐ Approve   Signature & Title
☐ Disapprove   ☐ Disapprove

☐ Approve   Signature & Title   ☐ Approve   Signature & Title
☐ Disapprove   ☐ Disapprove

☐ Approve   Signature & Title
☐ Disapprove

Remarks of Service/Awards Committee

_____
_____
_____
_____

Attachments

☐ Supplemental Reports   ☐ EMS Report   ☐ Injury Report   ☐ Other Documents

(Awards Form continued)

Facts continued

Awards Program

AWARDS CEREMONY

October 23, 2004
6:00 PM

First Baptist Church
Valley Ridge & Old Orchard

FIREFIGHTER'S PRAYER

When I am called to duty, God, wherever flames may rage,
Give me the strength to save a life, whatever be its age.

Help me to embrace a little child before it's too late,
Or save an older person from the horror of that fate.

Enable me to be alert to hear the weakest shout,
And quickly and efficiently to put the fire out.

I want to fill my calling and to give the best in me,
To guard my neighbor and protect his property.

And if according to your will I have to lose my life,
Bless with your protecting hand my loving family from strife.

AMEN

Special "Thanks" to *Bank of the West*, *Medical Center of Lewisville* and *Sysco Foods* for sponsoring our 2004 Annual Awards Ceremony.

(Awards Program continued)

Gene Carey
Mayor

Mike Nowels
Mayor Pro Tem

Greg Tierney
Deputy Mayor Pro Tem

Dean Ueckert
Coucilman, Place 4

Tim Blair
Councilman, Place 2

Rudy Durham
Councilman, Place 5

Claude King
City Manager

Richard Lasky
Fire Chief

Awards & Ribbons Committee
Assistant Chief Tim Tittle
Captain Greg Kohn
Captain Ken Swindle
Driver/Engineer Steve Carter
Firefighter/EMT Kyle Allen
Firefighter/Paramedic James Byers
Firefighter/Paramedic David Scott

Master of Ceremonies
Assistant Chief Tim Tittle

Invocation
Chaplain Benny Grissom

Welcome
Chief Richard Lasky

Introductions
Assistant Chief Darrell Brown

Dinner
Mayor Gene Carney

Year in Review
Chief Richard Lasky

Highlight Video

Presentation of Awards
Years of Service
Honorable Mentions
Award of Exemlary Action
Award of Commendation
Company Citation
Rookie Rirefigher of the Year
Firefighter of the Year
Paramedic of the Year
Officer/Employee of the Year

Closing
Assistant Chief Tim Tittle

Boat 168 In Service Announcement

"Stand-by for an announcement"   ALL CALL TONES

The "Lewisville Fire Department would like to announce the retirement of Boat 169, Asset #7018 and welcome aboard the new Boat 168.

We would like to wish Boat 168 "The Bryan Jarvis" and Boat 169 "The Michael Paul" and those assigned to them, a safe journey and the skills necessary to provide the best possible service to our visitors and the citizens of the City of Lewisville."

Placed into service this 17th day of March, 2004.

Boat Christening and Dedication

March 2004

Welcome and Introductions	(Chief Tittle)
Invocation	(Chaplain Grissom)
Chief Tittle	
Mayor Carey, City Council, City Manager	
Chief Lasky	
Blessing of the Fleet	(Chaplain Grissom)
Reveal the name of Boat 169	(Chief Lasky)
Reveal the name of Boat 168	(Chief Lasky)
Tones and Announcement	(Dispatch)
Bottle Breaking (Boat 168 "Named")	
Launch	
Conclusion	

Quint 162 in Service

"Lewisville Fire Department would like to announce the retirement of Quint 162, Asset #7812 and welcome aboard the new Quint 162, Asset #201123.

We would like to wish this Quint and those assigned to her a safe journey and the skills necessary to provide the best possible service to our visitors and the citizens of the City of Lewisville."

Placed into service this _____ day of _____, 2004.

Truck 161 in Service

"Lewisville Fire Department would like to announce the retirement of Truck 161, Asset #3100 and welcome aboard the new Truck 161, Asset #200449.

We would like to wish this Truck Company and those assigned to her a safe journey and the skills necessary to provide the best possible service to our visitors and the citizens of the City of Lewisville."

Placed into service this _____ day of _____, 2004.

Station 6 in Service

"Lewisville Fire Department would like to announce the opening of Firehouse #6, 2120 Midway Road and welcome aboard Engine 166, Asset #200449.

We would like to wish this Firehouse and those assigned to her a safe journey and the skills necessary to provide the best possible service to our visitors and the citizens of the City of Lewisville."

Placed into service this _____ day of _____, 2001.

Swearing-in Program

SWEARING IN CEREMONY

April 14, 2004
6:00 PM

Central Fire Station

"I have no ambition in the world but one, and that is to be a fireman. The position may, in the eyes of some, appear to be a lowly one; but we who know the work, which a fireman has to do believe that his is a noble calling. There is an adage which states that 'Nothing can be destroyed except by fire'. We strive to preserve from destruction the wealth of the world, which is the product of the industry of men, necessary for the comfort of both the rich and the poor. We are the defenders from fire, of the art, which has beautified the world, the product of the genius of men and the means of refinement of mankind. But, above all, our proudest endeavor is to save lives of men – the work of God himself. Under the impulse of such thoughts, the nobility of the occupation thrills us and stimulates us to deeds of daring, even at the supreme sacrifice. Such considerations may not strike the average mind, but they are sufficient to fill to the limit our ambition in life and to make us serve the general purpose of human society."

Edward F. Croker
FDNY
Chief of Department
(1899-1911)

(Swearing-in Program continued)

Gene Carey
Mayor

Dean Ueckert
Mayor Pro Tem

Mike Nowels
Deputy Mayor Pro Tem

Greg Tierney
Coucilman, Place 1

Tim Blair
Councilman, Place 2

Rudy Durham
Councilman, Place 5

Claude King
City Manager

Richard Lasky
Fire Chief

Welcome
Richard Lasky
Fire Chief

Introduction
Richard Lasky
Fire Chief

Administration of Oath
Mary Hendrix
City Secretary

Presentation of Badge and Certificates of Appointment
Gene Carey
Mayor

Firefighter
Gregory Rohre

Please join us for cake and coffee after the ceremony

Oath of Office SOP

Effective Date:_____

Approval:_____

Standard Operating Procedures

SECTION: **ADMINISTRATION**

TOPIC: **OATH OF OFFICE**

REFERENCE: **8.01**

I. OATH OF OFFICE

    A. All members will take the following oath during their "Swearing in" ceremony at initial hiring:

"I, _____, do solemnly swear, that I will faithfully execute the duties of the office of _____ of the City of Lewisville, State of Texas, and will, to the best of my ability, preserve, protect and defend the Constitution and Laws of the United States and of this State and the Charter and Ordinances of this City; and I furthermore solemnly swear, that I have not directly or indirectly paid, offered, or promised to pay, contributed, nor promised to contribute any money or valuable thing, or promised any public office or employment as a reward to secure my appointment. So help me God."

Oath of Office

CITY OF LEWISVILLE

COUNTY OF DENTON/DALLAS

STATE OF TEXAS

OATH OF OFFICE

"I, _____, do solemnly swear, that I will faithfully execute the duties of the office of **firefighter** of the city of _____, state of _____, and will to the best of my ability preserve, protect and defend the Constitution and Laws of the United States and of this state and the charter and ordinances of this city; and I furthermore solemnly swear, that I have not directly or indirectly paid, offered, or promised to pay, contributed, nor promised to contribute any money or valuable thing, or promised any public office or employment as a reward to secure my appointment. So help me God."

Signed _____

Sworn to and subscribed before me, this the _____ **day of** _____.

_____

Notary Public

Oath of Office Certificate

Statement of Appointment

I, _____, *do solemnly swear (or affirm), that I have not directly or indirectly paid, offered, promised to pay, contributed, or promised to contribute any money or thing of value, or promised any public office or employment, as a reward to secure my appointment or confirmation thereof, so help me God.*

Affiant's Signature
Division Chief for the Fire Department
Office to Which Appointed
City of Lewisville
City and/or County

SWORN TO and witnessed before me by affiant on this __th day of ___, 2005.

_____

Mayor

Promotional Program

PROMOTIONAL CEREMONY

August 19, 2004
6:00 PM

Central Fire Station

"I have no ambition in the world but one, and that is to be a fireman. The position may, in the eyes of some, appear to be a lowly one; but we who know the work, which a fireman has to do believe that his is a noble calling. There is an adage which states that 'Nothing can be destroyed except by fire'. We strive to preserve from destruction the wealth of the world, which is the product of the industry of men, necessary for the comfort of both the rich and the poor. We are the defenders from fire, of the art, which has beautified the world, the product of the genius of men and the means of refinement of mankind. But, above all, our proudest endeavor is to save lives of men – the work of God himself. Under the impulse of such thoughts, the nobility of the occupation thrills us and stimulates us to deeds of daring, even at the supreme sacrifice. Such considerations may not strike the average mind, but they are sufficient to fill to the limit our ambition in life and to make us serve the general purpose of human society."

Edward F. Croker
FDNY
Chief of Department
(1899-1911)

(Promotional Program continued)

Gene Carey
Mayor

Mike Nowles
Mayor Pro Tem

Gret Tierney
Deputy Mayor Pro Tem

Tim Blair
Councilman, Place 2

Dean Ueckert
Councilman, Place 4

Rudy Durham
Councilman, Place 5

Claude King
City Manager

Richard Lasky
Fire Chief

Welcome
Richard Lasky
Fire Chief

Introduction
Timothy Tittle
Assistant Chief

Administration of Oath
Julie Heinze
Deputy City Secretary

**Presentation of Badge and
Certificates of Appointment**
Rudy Durham
Councilman, Place 5

Assistant Chief
Darreli Brown

Battalion Chief
Jerry Wells

Captain
Donald Scarborough

Driver/Engineers
Michael Spinuzzi
Stephen Spraggins

Firefighters
Jason Martinson
Chris Croom
Michael Patrick
Kevin Pharr

Please join us for cake and coffee after the ceremony

Retirement Program

RETIREMENT CEREMONY

Division Chief Kenneth Wilkins

November 19, 2004
3:00 PM

Municipal Annex

I have no ambition in the world but one, and that is to be a fireman. The position may, in the eyes of some, appear to be a lowly one; but we who know the work, which a fireman has to do believe that his is a noble calling. There is an adage which states that 'Nothing can be destroyed except by fire'. We strive to preserve from destruction the wealth of the world, which is the product of the industry of men, necessary for the comfort of both the rich and the poor. We are the defenders from fire, of the art, which has beautified the world, the product of the genius of men and the means of refinement of mankind. But, above all, our proudest endeavor is to save lives of men – the work of God himself. Under the impulse of such thoughts, the nobility of the occupation thrills us and stimulates us to deeds of daring, even at the supreme sacrifice. Such considerations may not strike the average mind, but they are sufficient to fill to the limit our ambition in life and to make us serve the general purpose of human society."

Edward F. Croker
FDNY
Chief of Department
(1899-1911)

(Retirement Program continued)

Welcome
Carrell Brown
Introduction
Timothy Tittle

KENNY will always be a member of our family.

PLEASE JOIN US IN WISHING HIM
WELL ON HIS RETIREMENT

Please join us for cake and coffee after the ceremony

Gene Carey
Mayor

Mike Nowels
Mayor Pro Tem

Gret Tierney
Deputy Mayor Pro Tem

Dean Ueckert
Councilman, Place 4

Tim Blair
Councilman, Place 2

Rudy Durham
Councilman, Place 5

Claude King
City Manager

Richard Lasky
Fire Chief

Retired Member Radio Announcement

(Long Pager Tone) ALL STATIONS page

"All station companies stand by for a city-wide broadcast"

(Station 6 pager tones)

"Attention Engine 166, ID# 5264 (Retiring member's employee or badge number) be advised this is your last alarm from 2120 Midway Road., Station #6 in Box 632"

"The alarm is under control and tapped out at 14:35 hours"

"From the brotherhood and all the citizens in Lewisville, Thanks Captain Flanagan for the 34 years of service"

"KWF-636 Fire Department City of Lewisville"

LEWISVILLE FIRE DEPARTMENT

IS PROUD TO PRESENT THIS

CERTIFICATE

DECLARING

_____

Honorary Fire Chief

DATED THIS 7TH DAY OF JANUARY 2005

Richard A. Lasky, Fire Chief

September 11th — Annual Memorial Radio Announcement

0905 hours — "Stand by for an announcement."

>> 1 Alarm Tone

"All Stations, please observe a moment of silence in honor of those that lost their lives on September 11, 2001."

0928 hours – "Stand by for an announcement."

>> 1 Alarm Tone

"All Stations, please observe a moment of silence in honor of those that lost their lives on September 11, 2001."

Firefighter's Prayer

When I am called to duty, God wherever flames may rage,
Give me the strength to save a life, whatever be its age.

Help me to embrace a little child before it's too late,
Or save an older person from the horror of that fate.

Enable me to be alert to hear the weakest shout,
And quickly and efficiently to put the fire out.

I want to fill my calling and to give the best in me,
To guard my neighbor and protect his property.

And if according to your will I have to lose my life,
Bless with your protecting hand my loving family from strife.

– Author Unknown

Chief Croker Quote

"I have no ambition in the world but one, and that is to be a fireman. The position may, in the eyes of some, appear to be a lowly one; but we who know the work, which a fireman has to do believe that his is a noble calling. There is an adage which states that 'Nothing can be destroyed except by fire'. We strive to preserve from destruction the wealth of the world, which is the product of the industry of men, necessary for the comfort of both the rich and the poor. We are the defenders from fire, of the art, which has beautified the world, the product of the genius of men and the means of refinement of mankind. But, above all, our proudest endeavor is to save lives of men – the work of God himself. Under the impulse of such thoughts, the nobility of the occupation thrills us and stimulates us to deeds of daring, even at the supreme sacrifice. Such considerations may not strike the average mind, but they are sufficient to fill to the limit our ambition in life and to make us serve the general purpose of human society."

Edward F. Croker
FDNY
Chief of Department
(1899–1911)

A Fire Officer's Pledge

I promise concern for others.
A willingness to help all those in need.
I promise courage—courage to face and conquer my fears.
Courage to share and endure the ordeal of those who need me.
I promise strength—strength of heart to bear whatever burdens might be placed upon me.
Strength of body to deliver to safety all those placed within my care.
I promise the wisdom to lead, the compassion to comfort, and the love to serve unselfishly whenever I am called.

– *Author Unknown*

D
AFTER THE FIRE

AFTER THE FIRE

INSTRUCTION BOOKLET

Because we care

www.cityoflewisville.com

After the Fire Cover Letter

Dear Citizen:

The Lewisville Fire Department strives to serve the community of Lewisville by saving lives and property. Firefighters are familiar with the devastation and trauma resulting from fire. Generally, those who experience fire are not. The difficult period directly after a fire is confusing and traumatic. We recognize this and have created this booklet to assist you through this trying and tragic period.

While firefighters are on scene they will attempt to assist you and your family as best they can. After they leave questions may arise. By referring to this booklet, many of your concerns may be addressed. A list of important resources and telephone numbers are also included to speed your recovery from this unexpected event.

If we can help you in any other way, please do not hesitate to call the Lewisville Fire Department at (000) 123-4567.

Richard A. Lasky
Fire Chief
City of Lewisville

After the Fire General Information

TELEPHONE NUMBERS

American Red Cross	(000) 123-4567
Animal Control	(000) 123-4567
Auto Registration (County Tax Office)	(000) 123-4567
Building Inspection & Permits	(000) 123-4567
Chamber of Commerce	(000) 123-4567
Christian Community Action	(000) 123-4567
City Health Department	(000) 123-4567
City Main Switchboard	(000) 123-4567
County Courthouse	(000) 123-4567
Driver's Licenses (Dept. of Public Safety)	(000) 123-4567
Federal Information Center	(000) 123-4567
Fire Department – Administration	(000) 123-4567
Fire Department (Emergency)	911
Humane Society	(000) 123-4567
Internal Revenue Service	(000) 123-4567
Medical Emergencies	911
Police – Administration	(000) 123-4567
Police (Emergency)	911
Sanitation (Waste Management)	(000) 123-4567
Social Security	(000) 123-4567
Veterans Information	(000) 123-4567
Voters Information	(000) 123-4567
Welfare Office	(000) 123-4567

RECORDS AND DOCUMENTS

Records and documents are very important to your wellbeing and can be damaged or destroyed as a result of a fire. For this reason, the Lewisville Fire Department provides the following list of records and documents that should be located and/or replaced. Locating these documents will speed up the process of recovering from a fire.

Driver's License	(000) 123-4567
Checks	your bank
Insurance Policies	your insurance agent
Military Discharge Papers	your local service recruiter
Passports	(000) 123-4567
Marriage License	County Courthouse or state where ceremony was performed
Divorce Decree	County Courthouse or circuit court where decree was issued
Social Security Card	(000) 123-4567
Credit Cards	issuing companies
Titles to Deeds	County Courthouse or county where property is located
Stocks	issuing company or your broker
Wills	your lawyer
Medical Records/Prescriptions	your physician or pharmacist
Warranties	issuing company
Income Tax Records	(000) 123-4567
Auto Registration/Title	(000) 123-4567
Aging & Adult Services	(000) 123-4567
Welfare Office & Food Stamp Cards	(000) 123-4567

*Note: It is wise to store all important documents and records in an approved container that is specifically designed for such purposes.

IMPORTANT NAMES/TELEPHONE NUMBERS

Insurance Co. _____

Insurance Adjuster _____

Contractor _____

Plumber _____

Bank(s) _____

Doctor(s) Office _____

Dentist _____

Pharmacy _____

School(s) _____

Veterinarian _____

This publication was prepared by the City of Lewisville Fire Department as an aid to fire victims. There are no warranties made in connection with this publication, and the City of Lewisville shall not be held responsible for any damages (consequential, special or otherwise) arising from its use.

HOUSING ASSISTANCE

Contact your local disaster relief services agency such as the American Red Cross or the Salvation Army if you are in need of temporary housing. They have other services to help fire victims. Give them a call to see if they can assist you in any way.

If you are insured under a package homeowner's or tenant's policy, a section of your coverage may pay for temporary housing.

IF YOU MUST LEAVE . . .

This may be your decision or that of the Fire Department or building inspector if the building is unsafe. If you must leave:

- Contact the Police Department (000) 123-4567 so they can keep an eye on the property during your absence.
- Try to locate the following items to take with you:
    - All important identification
    - Vital medicines such as insulin or heart medication
    - Eyeglasses, hearing aids or other personal aids
    - Valuables such as money, insurance policies, credit cards, jewelry, checkbook, etc.

If you feel you will be out of your building for an extended period of time, you may want to notify the following of your relocation:

- Post Office to forward your mail to your new address
- Your bank(s)
- Utility companies
- Social Security Administration
- Insurance company
- Fire Department (if the fire is under investigation)
- Newspapers and magazines you may subscribe to

If there is structural damage to your building, check with the City building department to see if there is a need for a permit before attempting repairs.

INSURANCE

INSURED

Contact your insurance agent as soon as possible after a fire. If you are renting the property you must contact the owner as well. Your insurance agent may be able to help you in making immediate repairs or help in securing your home. If you cannot reach your agent and need professional assistance in boarding up your home, a general contractor or fire damage restoration firm can help. Check your Yellow Pages.

Remove as many valuables as possible if you must stay elsewhere. Be sure to inventory the property you remove. Also, check for important documents that may have been damaged.

Mobile home insurance coverage is similar to other forms of homeowners coverage. Check with your agent for the specifics regarding your coverage.

UNINSURED

If your property is not insured, or if your insurance will not cover all your losses, contact your family lawyer. You may have to depend on your own resources and help from other agencies to recover your fire loss.

The American Red Cross, Salvation Army, local church groups or civic organizations such as Rotary or Christian Community Action may be able to provide assistance.

Some losses due to fire are tax deductible for your federal income tax. Be sure to keep receipts for money spent on repairs or replacing damaged property and in covering your living expenses. These receipts will be helpful in calculating the loss for your yearly tax return.

Check with your local Internal Revenue Service office for PUBLICATION 547, TAX INFORMATION ON DISASTERS, CASUALTY LOSSES AND THEFTS. A quick refund is possible if you file Form 1045, APPLICATION FOR TENTATIVE REFUND. Check with the I.R.S. first.

MONEY REPLACEMENT

PAPER CURRENCY

NOTE: Handle burned money as little as possible. Attempt to encase each bill or portion of bill in plastic wrap for preservation.

If the money you've kept in your home is only half burned or less, you can check with any local commercial bank or take the remainder to the Federal Reserve Bank, or you can mail the remainder of the money (in plastic wrap) via first class mail to:

U.S. Treasury Department
Main Treasury Building, Room 1123
Washington, D.C. 20220

COINS

Mutilated or melted coins can be taken to the Federal Reserve Bank or mailed via first class registered mail to:

U.S. Mint
5th and Arch Street
Philadelphia, PA 19015

SAVINGS BONDS

If your U.S. Savings Bond(s) have been mutilated or destroyed, write to:

U.S. Treasury Department
Bureau of Loans and Currency
537 W. Clark Street
Chicago, IL 60605
ATTN: Bond Consultant

Be sure to include name(s) and address on bonds, approximate date or time period when purchased, denominations and approximate number of each.

SECURING THE SITE

The Fire Department will remove as much water and debris as possible from the fire building before turning the building over to the owner. It is the responsibility of the owner to see that the property is secure after the Fire Department leaves the scene. In the event the Fire Department feels the building is unsafe, we will secure the property as best we can.

CAUTIONS

Household wiring which may have been water damaged should be checked by an electrician before the current is turned back on.

The Fire Department will see that the utilities (water, electric, or natural gas) are either safe to use or are disconnected before we leave. The utility companies will not make repairs on the customer's side of the meter; therefore, a private contractor will have to be contacted to make repairs. All repairs of this nature require permits and inspection by proper building department personnel. The utility companies will not restore your utilities until the repairs are approved by the building department. <u>DO NOT ATTEMPT TO RECONNECT UTILITIES YOURSELF!</u>

Be watchful for any structural damage caused by the fire. The Fire Department will secure property we believe to be a safety hazard.

Any food or beverages that had contact with smoke, soot, or heat should be discarded. Wash your canned goods and jars in soapy water. If the labels come off, remark them with a permanent marker if you know the contents. Don't use canned goods when cans have bulges, are dented or show rust. IF YOU ARE UNSURE, THROW IT OUT!

Any medications that had contact with smoke, soot, or heat should also be thrown out. If you are uncertain about the reusability of the medication, DISCARD IT. Notify your physician and/or your pharmacist for replacement.

If your power has been turned off KEEP DOORS TO REFRIGERATOR AND FREEZER CLOSED! Refrigerators and freezers left unopened will hold their temperature for a short time. If your food becomes spoiled or thawed, THROW IT OUT! (For more information on saving foods, see salvage hints.)

If you have a safe, <u>DO NOT ATTEMPT TO OPEN IT!</u> Hot gases could burst into flames when the door is opened. Wait until the safe has cooled.

SALVAGE HINTS

The following salvage information was furnished by the Fire Center of the University of Minnesota as reprinted by the Federal Emergency Management Agency, U.S. Fire Administration.

These hints are meant as an economical way to clean up or salvage items after a small fire. Be sure to contact your insurance company to see exactly what they will cover. Also, consider taking pictures of the damage.

CAUTION

Several cleaning mixtures contain the substance Tri-Sodium Phosphate. Tri-Sodium Phosphate is a caustic substance used commonly as a cleaning agent. It should be used with care and stored out of the reach of children and pets. Wear rubber gloves when using it. Read the instructions on the container before you start. (Tri-Sodium Phosphate, also known as TSP, can be purchased in your local hardware, paint or home improvement store.)

- Vacuum all surfaces.
- Change and clean air conditioner filters.
- Seal off the room in which you are working with plastic to keep soot from moving from one location to another. Try to keep windows closed.

WALLS AND CEILINGS

To remove soot and smoke from painted walls, mix together four to six tablespoons of Tri-Sodium Phosphate and one gallon of water.

Wash a small area at a time, working from the floor up. Do ceilings last. Rinse thoroughly. <u>DO NOT REPAINT UNTIL COMPLETELY DRY!</u> It is advisable that you use a smoke sealer (available where paint is sold) before painting.

Wallpapered walls usually cannot be restored. Check with your wallpaper dealer.

CARPETS AND RUGS

Carpets and rugs should be allowed to dry thoroughly. Throw rugs can be cleaned by beating, sweeping or vacuuming, and shampooing. Rugs should be dried as quickly as possible. Lay them flat and expose them to a circulation of warm, dry air. A fan turned on the rugs will speed drying. Make sure the rugs are thoroughly dry. Even though the surface seems dry, moisture remaining at the base of the tufts can quickly rot a rug. For more information on cleaning and preserving carpets, call your carpet dealer.

MATTRESSES

Reconditioning an innerspring mattress at home is nearly impossible. Your mattress might be able to be renovated by a company that builds or repairs them.

If you must use your mattress temporarily, put it out in the sun to dry, then cover it with plastic sheeting. It is impossible to remove the odor of smoke out of pillows. The foam and feathers hold the odor in.

WOOD FURNITURE

Do not use chemicals on furniture. A very inexpensive product called FLAX SOAP (available in hardware and paint stores) is a most efficient product to use on wood, including kitchen cabinets. If you do not have Flax Soap:

- Wipe off with Borax dissolved in hot water to remove mildew.
- To remove white spots or film, rub the surface with a cloth soaked in a solution of ½ water and ½ vinegar. Then wipe dry and polish with wax.
- You can also rub the wood surface with steel wool in liquid polishing wax, wipe with soft cloth and then buff.

NOTE: Wet wood can decay and mold, so dry well **BUT DO NOT DRY IN DIRECT SUNLIGHT** as the wood may warp and twist out of shape.

WOOD AND VINYL FLOORS

Use Flax Soap on wood and vinyl floors. It will take 4 to 5 applications. Then strip and rewax. When water gets underneath linoleum, it can cause odors and warp the floor. If this has happened, remove your linoleum. Call your linoleum dealer for suggestions on how to loosen the adhesive without damaging the floor covering. Be sure to let the floor dry thoroughly before replacing it.

LOCKS, HINGES, TYPEWRITERS AND SMALL APPLIANCES

Steam from a fire removes lubrication from these items. They should be taken apart and oiled. It is suggested that these items be taken to a repair shop. If locks cannot be removed, squirt machine oil through a bolt opening or keyhole and work the knob to distribute the oil. Hinges should also be thoroughly cleaned and oiled.

COOKING UTENSILS

Your pots, pans, flatware, etc., should be washed with a fine-powdered cleanser. You can polish copper and brass with special polish, or salt sprinkled on a cloth saturated in vinegar.

REFRIGERATORS AND FREEZERS

To remove odors from your refrigerator or freezer, wash the inside with a solution of baking soda and water or use one cup of vinegar or household ammonia to one gallon of water. Some baking soda in an open container, or a piece of charcoal can be placed in the refrigerator or freezer to absorb odor.

> **Caution:** When cleaning or discarding any refrigerator or freezer, be sure the doors are removed or secured against closing on a young child.

FOOD

If your freezer has stopped running, you can still save the frozen food:

- Keep the freezer closed. Your freezer has enough insulation to keep food frozen for at least one day, perhaps more.
- Move your food to a neighbor's freezer or a commercial freezer firm. Wrap the frozen food in newspaper and blankets, or use insulated boxes.

If your food has thawed, observe the following procedures:

- FRUITS can be refrozen if they still taste and smell good. Otherwise, if the fruits are not spoiled, they can be eaten.
- VEGETABLES should not be refrozen if they have thawed completely. Refreeze only if they have ice crystals in the vegetables. If your vegetables have thawed and cannot be used soon, THROW THEM AWAY! If you have any doubts whether your vegetables are spoiling, THROW THEM AWAY! Don't wait for a bad odor.
- MEATS may be refrozen (if the ice crystals remain). Cook very thoroughly before eating the meat. If odor is poor or if you question these foods, THROW IT AWAY, as bacteria multiply rapidly.

CLOTHING

Smoke and soot can sometimes be removed from clothing. The following formula will often work for clothing that can be bleached:

4-6 teaspoons Tri-Sodium Phosphate
1 c. Lysol or any household chlorine bleach
1 gallon warm water
Mix well—add clothes, rinse with clean water—dry well

To remove mildew, wash the fresh stain with soap and water. Then rinse and dry in the sun. If the stain isn't gone, use lemon juice and salt, or diluted solution of household chlorine bleach. TEST COLORED GARMENTS BEFORE USING ANY TREATMENT! Take wool, silk or rayon garments to the dry cleaners as soon as possible.

LEATHER AND BOOKS

Wipe your leather goods with a damp cloth, then with a dry cloth. Stuff your purses and shoes with newspapers to retain their shape. Leave your suitcases open. Leather goods should be dried away from heat and sun. When leather goods are dry, clean with saddle soap. You can use steel wool or a suede brush on suede. Rinse leather and suede jackets in cold water and dry away from heat or sun.

Books can be dried by standing them up with their pages separated. Then they should be piled and pressed to prevent the pages from crinkling. Alternating drying and pressing will prevent mildew from forming until the books are thoroughly dry. If your books are very damp, sprinkle cornstarch or talc between the pages, leave for several hours, then brush them off. A fan turned on the books will help them dry.

HOW TO USE 9-1-1

1. 9-1-1 can be dialed from any phone in the County area. No change is needed when using a pay phone.

2. When calling, state the nature of the emergency and the address where emergency aid is needed.

3. Give the address of the emergency and the number from the phone you are using.

4. REMAIN CALM and answer any questions that the 9-1-1 operator has for you. The operator wants to help you, but won't be able to if you are too excited.

5. Speak clearly and do not shout into the phone.

6. Do not hang up the phone until the 9-1-1 operator has done so.

FIRE PREVENTION TIPS

1. Install and maintain smoke detectors. These are best installed on the ceiling adjacent to sleeping areas.

2. Avoid overloading electrical outlets. This condition can cause fires due to an excessive amount (3 or more) of appliances in an outlet.

3. Have an escape plan for every person in the house. Two escape routes, either through a door or window, are recommended per room.

4. Have a meeting place to go to if escaping a fire. For the children's sake, call it "home base" or "safe zone" because they respond to this as well as identify with the concept. This should be a place away from the home, preferable a mailbox, neighbor's porch, etc.

5. If a fire occurs get out! Call the Fire Department. Fires grow quickly! A small fire can, in less than seconds, become a raging, choking inferno. Leave it to the professionals. Also, often there will be less damage than if you and neighbors attempt extinguishment.

6. Post your address and telephone number by the phone for babysitters, guests, and even yourself to use in case of emergency. 9-1-1 is uncomplicated; however, you may even forget your own name and address under extreme stress.

7. If a pan fire occurs, put a lid or cover on it. Tip the lid so it shields your arm from the flames as you place it. Turn off the heat. AVOID water or application of any towel, wet or dry, on a pan fire.

Call the Fire Department!

HOW YOU CAN HELP WHEN EMERGENCY PERSONNEL ARE AT YOUR HOME

1. If it is a fire, stay away from the house or apartment on fire.

2. Tell firefighters if there are any other persons in the structure.

3. If an animal is still in the structure, tell the firefighter that an animal is inside the home. Avoid saying only the animal's name or a special term used for the animal. This leads to confusion, resulting in firefighters looking for a human.

4. Remain as calm as possible. Stay with friends or other family members, but please do not leave the fire location unless to accompany a rescue vehicle.

5. If you accompany an ambulance to a hospital via a private car, DO NOT speed or go through any red lights. The ambulance drivers are trained, licensed, and signal equipped for this. Another accident on the way to the hospital will not help a victim. Remember—they are in qualified, caring hands.

6. If it is an ambulance call, state as clearly as possible any medications or medical problems the patient might have. If you do not know the answers, say so! You are helping speed care by being clear and truthful.

7. If you have been directed not to enter a structure after a fire, heed this warning. Remaining smoke is a killer, along with the possibility of structural weakness in the building.

8. As much as you may wish to physically assist emergency personnel, please avoid the temptation. This is a team effort at your service. Only a limited number of specially trained persons can make it work. Follow the previous steps, stay on the fringe, and we will assist you in the caring, professional manner that we are equipped for.

ABOUT OUR OPERATIONS

Here are a few common questions people have about our operations.

WHY ARE WINDOWS BROKEN OR HOLES CUT IN THE ROOF?

As a fire burns, it moves upward and then outward. Breaking the windows and/or cutting holes in the roof (called ventilation) stops that damaging outward movement and enables firefighters to fight the fire more efficiently, resulting in less damage to the structure in the long run.

WHY DO WE CUT HOLES IN WALLS?

We have to be absolutely sure that the fire is completely out, and that there is no fire inside the walls or other hidden places. We will do the least amount of damage necessary to ensure everything is safe.

WHY DOES A LEWISVILLE FIRE ENGINE RESPOND TO MEDICAL EMERGENCIES WHEN THERE IS NO FIRE?

This allows faster response to an emergency, better use of manpower, and most importantly, expeditious care to patients. This program allows the nearest engine to quickly respond to a medical emergency, yet is always backed up by an advanced life support unit.

IS IT POSSIBLE TO OBTAIN A COPY OF THE FIRE REPORT?

Yes. A fire report is a public document and is available at the Lewisville Fire Department Administrative Office, located at 188 N. Valley Parkway. You can reach us by phone at 000-123-4567. Important information concerning your fire can be found on the back cover of this booklet.

A final note—The Lewisville Fire Department is here for you. It is made up of over 100 state certified firefighters. All paramedics and EMTs on the ambulances are certified firefighters. There are five stations throughout the City for your protection. Both advanced life support and basic life support are available. This allows for direct communication with hospitals and physicians during rescue. Full drug and advanced life saving techniques are available at your call.

We are here to serve. Our greatest concern and commitment is for the safety and well-being of the citizens of Lewisville.

If you have any questions or comments about any part of our operations, call the Fire Department.

Date_____

Time_____

Incident No. _____

Address of Fire_____

For additional information please call Fire Department Administration at (000) 123-4567.

Customer Support Unit Check Sheet
SUPPORT 160
1996 INTERNATIONAL - ASSET # 7299
CHECK SHEET

RETURN TO CHIEF   DATE:

	SUN	MON	TUE	WED	THU	FRI	SAT
FUEL							
OIL LEVEL/PRESSURE							
RADIATOR – WATER/COOLANT							
BATTERIES – CHARGE INDICATOR							
TRANSMISSION – FLUID LEVEL							
BRAKES							
BELTS/HOSES							
VEHICLE LIGHTS							
ALL EMERGENCY LIGHTS							
SIREN/HORN							
TIRES – WEAR/INFLATION							
OPTICOM							
RADIO CHECK (MOBILE)							

CHECKED BY DE: (INITIAL/DATE)							
CHECKED BY FF: (INITIAL/DATE)							
CHECKED OFF BY CAPT. (INITIAL/#)							

INSPECTED BY: _____   DATE: _____

OIL – QUARTS ADDED/CHANGED							
COOLANT GALLONS ADDED (#G)							
POWER STEERING FLUID (#QTS)							
TRANSMISSION FLUID (#QTS)							
CHARGE BATTERY/ADD WATER							

FLUIDS ADDED BY:							

COMMENTS
_____
_____
_____
_____
_____

SUPPORT 160
1996 INTERNATIONAL - ASSET # 7299
MONDAY CHECKLIST
FOR VEHICLE

RETURN TO CHIEF

HOURS/MILEAGE	
BRAKE FLUID LEVEL	
POWER STEERING FLUID LEVEL	
AIR FILTER CHANGE INDICATOR	
TRANSMISSION FLUID	
DIFFERENTIAL & TRANSFER FLUID LEVEL	
TIRES – WEAR/INFLATION	
LIST TIRE PRESSURES 1. _____ 4. _____ 2. _____ 5. _____ 3. _____ 6. _____	
STATE INSPECTION DUE:	

HOURS: _____   MILEAGE: _____

FRONT & REAR RANGE

90 TO 100 P.S.I.

☐ ☐

☐ ☐ ☐ ☐

DRIVER'S SIGNATURE: _____   DATE: _____

COMMENTS

_____
_____
_____
_____
_____
_____
_____
_____
_____
_____
_____
_____
_____

SUPPORT 160
1996 INTERNATIONAL - ASSET # 7299
CHECK SHEET

RETURN TO CAPTAIN

	SUN	MON	TUE	WED	THU	FRI	SAT
CABINET 1							
2 FLAT HEAD SHOVELS							
2 SPADE SHOVELS							
2 SQUEEGEES 24"							
5 T-POSTS							
4 HAND BROOMS							
CABINET 2							
2 SETS SAW HORSE LEGS							
1 CABLE BAILING WIRE							
3 ORANGE 5 GALLON BUCKETS							
POLYETHELENE – BLACK 20' X 100'							
GARBAGE SCOOP							
BAG OF STYROFOAM							
CABINET 3							
BOX TAPE							
EXTRA BLACK MARKERS							
EXTRA DUCT TAPE							
1 BOX YARD BAGS							
2 BOXES ZIP LOCK BAGS							
CABINET 4							
42 EACH 11' X 15" X 25" BOXES							
21 TARPS							
ORANGE CLEANER							
CABINET 5							
2 GAS CANS							
1 OIL CAN							
9 EA ½ " X 4' X 8' PLYWOOD							
12 EA 1" X 2' X 8' WHITE PINE							
1 EA 1" X 4' X 8' WHITE PINE							
1 10' LADDER							
10' X 10' EASY-UP AWNING							
2 EA 14 GAL 5.4 HP RIGID SHOP VAC							
12 EA 2 X 4 X 8 STUDS							
5500 HP GENERATOR							
C1500 TWIN HEAD LIGHTS W/CASE							
STREAMLIGHT HAND LIGHT							
1 SHOP VAC							
CORDLESS SCREWDRIVER/SAW--							
FLASHLIGHT WITH BAG							
CABINET 6							
¼" 100' AIR HOSE							
3/8" 100' AIR HOSE							
AIR COMPRESSOR							
SPOT NAILS							
PORTER CABLE NAIL GUN							

	SUN	MON	TUE	WED	THU	FRI	SAT
CABINET 7							
12 GAUGE 100; EXTENSION CORD W/3' GFI CORD ATTACHED							
12 GAUGE 25' EXTENSION CORD							
12 GAUGE 25' EXTENSION CORD W 3' GFI CORD ATTACHED							
12 GAUGE 25' EXTENSION CORD							
CABINET 8							
DEWALT CIRCULAR SAW							
DEWALT RECIPROCAL SAW							
148 PIECE TOOL SET/DRILL BITS							
MILK CRATE – PLASTIC							
2 YELLOW SCREW/NAIL HOLDERS							
SCREW HOLDER ASSORTED							
TOOL BOX #19							
TOOL BOX #20							
TOOL BOX #21							
AMMO BOX W/ 1-5/8" GALV NAILS							
AMMO BOX W/ 1-5/8" WOOD SCREWS							
2 EACH 36" CROW BAR							
18" CROW BAR							
AMMO BOX W/3-1/2" GALV NAILS							
AMMO BOX W/1-1/4" GALV NAILS							
8D 2-1/2" NAILS – 5 LB BOX							
CABINET 9							
HUB CAP TOOL							
TOOL BOX #22							
2 CLOTH NAIL APRONS							
2 LEATHER NAIL APRONS							
2 PLASTIC MILK CRATES							
CABINET 11							
14 OZ REGULAR SHAVING CREAM							
14 OZ SENSITIVE SHAVING CREAM							
BABY POP-UPS 80 CNT SCENTED							
BABY POP-UPS 80 CNT UNSCENTED							
8 CREST TOOTHPASTE (.85 OZ)							
42 DRYPERS, 16-28 LBS							
6 ORAL-B DENTAL FLOSS							
9 ORAL-B TOOTHBRUSH, ADULTS							
12 ORAL-B TOOTHBRUSH, CHILD							
30 PREEMIES DIAPERS, INFANT							
30 TWIN BLADE DISP. RAZORS							
CABINET 12							
PAPER TOWELS							
CABINET 13							
4 STORAGE STACKERS							
ASSORTED GLOVES							
GOGGLES							
PLASTIC DIVIDERS							
4 HARD HATS							

	SUN	MON	TUE	WED	THU	FRI	SAT
CABINET 14							
BLACK MARKERS							
DANGER TAPE ROLLS							
TAPE-DUCT AND ELECTRICAL							
CABINET 16							
HINGES							
LATCHES							
LOCKS							
3 EA BLACK SPRAY PAINT							
NUTS, BOLTS AND SCREWS							
CABINET 17							
12 DISP. THERMAL BLANKETS							
CABINET 18							
GOGGLES							
CAB							
2 GAS KEYS (DIESEL & GASOLINE)							
MAPSCO – DALLAS							
MAPSCO – DENTON COUNTY							
MAPSCO – FT WORTH							
STREAMLIGHT HAND LIGHT							
CHECKOUT BOOK							
CONSUMABLES BOOK							
LOG BOOK							
TOOLBOX #19 (IN CABINET NUMBER 8)							
3 EA 25' TAPE MEASURE							
4 LB DRY WALL HAMMER							
2 CARPENTER HAMMER							
SHARP TOOTH HAND SAW 15"							
3 PHILLIPS SCREWDRIVERS							
4 FLAT HEAD SCREWDRIVERS							
CARPENTER PENCIL SHARPENER							
CARPENTER PENCILS							
2 PKS SCRATCH PADS – 4							
18' PRY BAR							
HACK SAW							
3 - HACK SAW BLADES (IN PVC PIPE)							
2 VICE GRIPS							
2 UTILITY KNIVES W/BLADES							
3 CHANNEL LOCK PLIERS							
STUD FINDER							
3 WOOD CHISELS							
CHALK LINE							
SLEDGE HAMMER							
CRESCENT WRENCH							
TIN SNIPS							
TORPEDO LEVEL							
DRY WALL KNIFE (JAB SAW)							
MAGIC MARKER							

	SUN	MON	TUE	WED	THU	FRI	SAT
TOOLBOX #20 (IN CABINET NUMBER 8)							
CLAW HAMMER							
WOOD SAW							
HOOKS							
DRYWALL HAMMER							
TOOLBOX #21 (IN CABINET NUMBER 8)							
LEVEL – 18"							
24" BOLT CUTTERS							
TOOLBOX #22 (IN CABINET NUMBER 9)							
DEWALT DRILL							
DEWALT DRILL BIT CASE							
14 PC PADDLE BIT SET							

COMMENTS

_____
_____
_____
_____
_____
_____

SUPPORT 160
1996 INTERNATIONAL - ASSET # 7299
MONDAY CHECKLIST
FOR EQUIPMENT

RETURN TO CAPTAIN

CHAIN SAW FUEL		
CHAIN SAW BAR OIL		
CHAIN SAW START		
BLOWER FUEL		
BLOWER START		

INSPECTED BY: _____   DATE: _____

COMMENTS

_____
_____
_____
_____
_____
_____

Miscellaneous Forms

**CITY OF LEWISVILLE
FIRE DEPARTMENT
RECORD OF EXCEPTIONAL PERFORMANCE**

1. Employee: _____ 2. Division/Section: _____

3. Classification: _____ 4. Date Prepared: _____

5. Initiator of commendation: _____

6. Description and date of exceptional performance:
_____
_____
_____
_____

7. Supervisor's comments and date of presentation:
_____
_____
_____

8. Employee's comments:
_____
_____
_____

9. Supervisor's signature   10. Employee's signature

_____   _____

Original: FD Employee File
Copy: Employee
            Fire Chief
            Personnel Files, Human Resources

REQUEST TO RIDE OUT WITH LEWISVILLE FIRE DEPARTMENT

PLEASE PRINT LEGIBLY

Name: _____ DOB: _____

Address: _____ Phone: _____

Emergency Contact: _____ Phone: _____

Dates/Times Requested for Ride Out:
1. _____
2. _____
3. _____

How would riding out with the Lewisville Fire Department benefit you?

_____

_____

_____

☐ Approved   ☐ Denied   (LFD USE ONLY)

_____   _____
Fire Chief (or Designee)   Date

RELEASE, ASSUMPTION OF RISK AND INDEMNIFICATION AGREEMENT FOR PARTICIPATION IN CITY OF LEWISVILLE FIRE DEPARTMENT TRAINING ACTIVITIES:

In consideration for the acceptance of my participation in City of Lewisville Fire Department training activities, including, but not limited to, riding in City of Lewisville apparatus, riding in a City of Lewisville ambulance as an assistant or observer, and the use by me of any self-contained breathing apparatus cylinder which may have been filled by the City of Lewisville Fire Department, and with the understanding that my participation is only on condition that I enter into this Agreement for me, my heirs, and assigns, I hereby assume the inherent and extraordinary risks involved in the activities and all other risks, inherent in any other activities in which I may voluntarily participate. I expressly assume the risk of, and accept full responsibility for, any and all injuries, including death, and accidents which may occur as a result of my participation and release from liability the City of Lewisville, each of its officers, directors, agents, representatives, employees and all other persons and entities associated with the training activities. I hereby waive any claim I may hereafter have as a result of any and all injuries to my person or property as a result of my participation in any activities connected with the City of Lewisville Fire Department in which I may participate. I hereby agree to indemnify all of the above named persons, for any and all claims, including attorney's fees and costs, which may be brought against any of them by anyone claiming to have been injured or damages as a result of my participation in the activities.

I CERTIFY THAT I HAVE READ AND FULLY UNDERSTAND THIS RELEASE. I AM OF LAWFUL AGE AND LEGALLY COMPETENT TO MAKE THIS AGREEMENT.

_____   _____
Signature   Date

_____   _____
Parent/Guardian Signature, if a minor   Date

LEWISVILLE FIRE DEPARTMENT / MEDIA INFORMATION

Incident Address: _____

Apt / Units / Exposures / Acres / Vehicles: _____

_____

Date: _____ Time Disp: _____ Time Arrived (A): _____

Tapped Out (B): _____ Total Time to Bring Under Control (A+B): _____

How Call Was Received: 9-1-1: _____ Passerby: _____ Radio: _____ Other: _____

LFD Alarms: _____ # of LFD Units: _____ # of LFD Firefighters: _____

Assisting Agencies: _____

_____ # of Firefighters: _____

Dollar Loss: Structure: $_____ Contents: $_____ Vehicle: $_____

Working Smoke Detector: YES / NO   Working CO Detector: YES / NO

Cause & Origin: Undetermined: YES / NO

Determined: _____

# of Families Displaced: _____ After The Fire: YES / NO   Red Cross: YES / NO

of: Injured: _____ Fatalities: _____ FFs Injured: _____ Pets: _____

Victim Status: #1: _____ #2: _____ #3: _____ #4: _____

#5: _____ #6: _____ #7: _____ #8: _____ #9: _____

MVA: Extrication Time: _____ Transported to: _____

Haz-Mat: Materials Involved: _____

_____

Additional Information: _____

_____

_____

LFD / MEDIA Checklist Form 2004

REQUEST FOR DONATIONS

Date: _____

Submitting Department: _____

Contact Name: _____

Contact Extension: _____

Approval: _____

Purpose of Donations: _____
_____
_____
_____
_____

Timeframe for Soliciting: _____

Businesses Contacted	Items or $0.00 Solicited

MAYDAY MAYDAY MAYDAY

UNKNOWN COMPANY / MEMBER MISSING

____ Announce Emergency Traffic

____ Declare a MAYDAY

____ PAR – Interior Companies First
(Dispatch to Verify Companies
On-Scene / en route via CAD Log)

____ What Company(s) / Member(s) are Missing, Lost or Trapped?

____ How Many Firefighter(s) are in the Missing Company(s)?

MISSING COMPANY / MEMBER IDENTIFIED

____ Announce Emergency Traffic

____ Declare a MAYDAY

____ Announce Who is Missing

____ Where was the Firefighter(s) Last-Known Location / Assignment?

____ Assign the RIT **(Remind All of Radio Discipline)**

____ Call for Next Greater Alarm Level **(Get Enough HELP to the Scene)**
(Can Interior Companies Hold Their Position?)

____ Request Additional EMS / Ambulance(s)

____ Building / Fire Conditions (Collapse Potential?)

____ Assign a Rescue Sector Officer **(Remind All of Radio Discipline)**
(Build a "Think Tank" for Command)

____ Assign Companies to Assist in **RIT SUPPORT**
(Also Consider Additional RIT Teams)

____ Conduct PAR of **ALL** Companies to Confirm Number of Missing

RETURNING TO NORMALCY

____ PAR

____ Reestablish a Plan of Attack for the Original Fire

____ Assess the Need for Additional Companies

____ Need for CISM?

MAYDAY MAYDAY MAYDAY

LFD / MAYDAY Checklist Form 2004

RAPID INTERVENTION TEAM CHECKLIST

ARRIVAL

____ Check in with Incident Command
____ Location of RIT

SIZE-UP

____ Building Occupancy
____ Building Construction Type:
  __ Wood Frame
  __ Heavy Timber
  __ Ordinary
  __ Noncombustible
  __ Fire Resistive
____ Placement of Windows / Doors / Fire Escapes / Porches / Egresses
____ Danger of High-Security Doors / Barred Windows / Building Modifications
____ Basement / Walkouts

TACTICS

____ Offensive / Defensive / Defensive-to-Offensive
____ Command Operations:
  __ Location of Command Post
  __ Communications / Frequency Used
____ Ladders and Truck Operations (Consider Ground Ladders to Upper Floors)
____ Fireground Time vs. Progress

EQUIPMENT

____ Stage Equipment Based on Construction Type

Wood Frame / Heavy Timber / Ordinary
____ RIT Tarp
____ Halligan Bar / Sledgehammer
____ Search Rope
____ Emergency Air Supply / SCBA
____ Pickhead Axes / Pike Poles
____ Ground Ladder(s)
____ Chain Saw / Circular Wood Saw
____ Circular Metal-Blade Saw
____ Oxygen / Defibrillator

Noncombustible / Fire Resistive
____ RIT Tarp
____ Halligan Bars
____ Sledgehammers
____ Search Ropes
____ Emergency Air Supply / SCBA
____ Ground Ladder(s)
____ Circular Metal-Blade Saw
____ Torch
____ Oxygen / Defibrillator

OTHER OPERATIONS

____ Potential Collapse / Collapse Area
____ Check with Safety Officer / Compare Information
____ Relocate and/or Add Another RIT
____ EMS for the RIT

LFD / RIT Checklist Form 2004